Growth and Maturation

AN INTRODUCTION
TO PHYSICAL DEVELOPMENT

Melvyn J. Baer

THE UNIVERSITY OF MICHIGAN

Howard A. Doyle *Publishing Company*

Cambridge, Massachusetts 02139

GROWTH AND MATURATION

Library of Congress Catalog Card Number 72-81296

To

Anna Baer

and to the memory of

Ben Baer

TABLE OF CONTENTS

Preface

From a biological point of view, a person moves insensibly from one level of development to another without discontinuity or break. Thus development is like a motion picture and the process must be seen in one sweep from beginning to end. The concepts of growth and maturation are here applied to the study of the skeleton, height and weight, body composition, the teeth, and the role of the endocrine glands. Hereditary and environmental factors and differences in the rate or timing of development are used to explain individual variation. Rather than relying on quantitative norms as arbitrary yardsticks for assessing the individual child, the emphasis is placed on understanding the pattern of development and the biological basis of individual variation.

This book introduces a series of working concepts in physical development, stressing the processes of growth and maturation. Specific facts are used only to illuminate and illustrate how these processes bring about the changes we observe in the human organism from conception through old age. The specialist in physical development may be critical of the absolute manner in which many generalizations have been stated. A number of the generalizations, viewed from a strictly technical position, require extensive qualification. To include these qualifications in an introductory text, however, often denudes the basic principle of its vitality and obscures the major point for the reader.

This book is addressed primarily to students and faculty in home economics, nursing, child development and family relations, and education. It attempts to incorporate significant technical ideas in a nontechnical way. The basic concepts of physical development are included without burdening the reader with complicated terms and references. The aim is to provide an account of how we grow and how we mature. If the reader comprehends a series of key ideas about the nature of physical development, the purpose of this book will have been achieved.

Acknowledgments

Many people contributed directly to the creation of this book. I am particularly indebted to Betti Kurtzman, Donna Harris, and Muriel Wagner, who participated in the production of a series of television programs on physical development sponsored by the Merrill-Palmer Institute. The series provided a springboard for this book. I would like to thank Ruth E. Cressman, Susan I. Seger, Jerri L. White, and Barbara A. Gruschow, librarians at The University of Michigan School of Dentistry, for locating information regarding bibliographic references, and Helen Konapek and Janice Everett for secretarial assistance. I am especially grateful to Samuel Damren for his continual help in acquiring materials and in the preparation of various sections of the manuscript. Doctors James E. Harris, Don B. Clewell, and Vincent C. Hascall were kind enough to read the manuscript critically, make suggestions regarding accuracy and ambiguity, and save the author from making a number of technical errors.

I wish to thank my wife, Bernice, for her patience and fortitude during her retyping of the many revisions of the manuscript. Finally, I must thank Clark Moustakas for encouraging me to write this book and for his unrelenting insistence that I complete it.

The many people who contributed illustrations for the text or who permitted reproduction of published drawings, diagrams, and X-rays are acknowledged below. The page on which the illustration appears is indicated.

M.J.B.

INDEX TO ILLUSTRATIONS

Illustration on page:

CHAPTER 1

CHAPTER 2

CHAPTER 3

Illustration on page:

CHAPTER 4

50, 69 (bottom)	From Stratz, C. H.: *Naturgeschichte des Menschen,* 1904. Verlag von Ferdinand Enke, Stuttgart.
51	© Copyright 1953, 1962 CIBA Pharmaceutical Company, Division of CIBA-GEIGY Corporation. Reproduced, with permission, from THE CIBA COLLECTION OF MEDICAL ILLUSTRATIONS by Frank H. Netter, M.D. All rights reserved.
52, 56 (bottom)	Drawn by Sheryl Morgan.
53, 54, 55, 56 (top)	Photographs by Bettie Shields Wagner.
57, 59, 66, 70, 71, 72, 73	Drawn by Dr. Gary C. Hollman.
62-63, 64, 65	From "Skeletal Maturation" by Dr. Hugo Mackay, published by Radiography Markets Division, Eastman Kodak Company.
67	Drawn by Rex Lamoreaux.
69 (top)	From unpublished thesis of Harry A. Wilmer (1940). Reprinted by permission of Burgess Publishing Company, from *Outline of Physical Growth and Development,* 1941, by Edith Boyd.
74	Drawn by Dr. Marvin Davis.
75	From Nigel Bateman: Bone Growth: A study of the grey-lethal and microphthalmic mutants of the mouse, *Journal of Anatomy. 88*: 212-262, 1954. Published by Cambridge University Press.

CHAPTER 5

80, 84	Reprinted from RADIOGRAPHIC ATLAS OF SKELETAL DEVELOPMENT OF THE HAND AND WRIST, SECOND EDITION by William Walter Greulich and S. Idell Pyle with the permission of the publishers, Stanford University Press. Copyright ©, 1950 and 1959 by the Board of Trustees of the Leland Stanford Junior University.
83, 86, 95	Drawn by Dr. Gary C. Hollman.
85	From Bayley, Nancy, and Pinneau, Samuel R.: Tables for predicting adult height from skeletal age: Revised for use with the Geulich-Pyle hand standards, *J. Pediat. 40*: 423-441, 1952.
88, 89	Reprinted by permission of The Society for Research in Child Development, Inc., from Shuttleworth, F. K.: Sexual maturation and the physical growth of girls age six to nineteen. *Monographs of the Society for Research in Child Development, 2,* Ser. No. 12, 1937.

Illustration on page:

91	From Garn, S. M., F. N. Silverman, K. P. Hertzog, and C. G. Rohmann: Lines and bands of increased density. *Medical Radiography and Photography, 44*: 58-89, 1968. Published by Radiography Markets Division, Eastman Kodak Company.
92	From Weinmann, Joseph P. and Sicher, Harry: *Bone and Bones,* ed. 2, St. Louis, 1955, The C. V. Mosby Co.; after H. R. Schurz, W. Baensch, and E. Friedl.
93	Radiograph courtesy of Prof. Albert G. Richards, The University of Michigan.
96	Drawn by Rex Lamoreaux.

CHAPTER 6

102	From Krogman, Wilton M., THE HUMAN SKELETON IN FORENSIC MEDICINE, 1962. Courtesy of Charles C Thomas, Publisher, Springfield, Illinois.

CHAPTER 7

110	From Harris, J. A., et al. *The Measurement of Man,* University of Minnesota Press, Minneapolis, © 1930. The University of Minnesota.
111	From unpublished thesis of Harry A. Wilmer (1940). Reprinted by permission of Burgess Publishing Company, from *Outline of Physical Growth and Development,* 1941, by Edith Boyd.
112, 118, 119	Drawn by Rex Lamoreaux.
113	From Reynolds, Earle L. and Patricia Grote: Sex differences in the distribution of tissue components in the human leg from birth to maturity, *The Anatomical Record. 102*: 45-53, 1948.
117	Drawn by Dr. Gary C. Hollman.
121, 122	From Keys, Ancel, et al. *The Biology of Human Starvation,* University of Minnesota Press, Minneapolis © 1950. The University of Minnesota.

CHAPTER 8

126	Adapted from the Baldwin-Wood Tables for Boys and Girls of School Age, 1923.
127 (top), 132	Drawn by Rex Lamoreaux.
127 (bottom), 131	Drawn by Dr. Gary C. Hollman.
129, 130, 133, 134, 136	Height-age and weight-age growth charts for boys and girls, ages: birth through six years, and five years through eighteen years. Copyright © 1943 by State University of Iowa.

Illustration on page:

Chapter 1

CULTURAL IDEAL OR
BIOLOGICAL REALITY?

The biologist's conception of development is vastly different from
that of the man in the street. While the average person may respond
to the artistic dimensions of physical appearance, the biologist
considers this a cultural bias. In his assessment of physical development, the biologist is not concerned with aesthetic qualities;
he is interested in studying changes in anatomical features through
time and in the emergence and metamorphosis of new tissues, organs,
and structures. The biologist seeks to establish a definition of development that has permanence and universal validity. In contrast,
aesthetic conceptions of development are transitory and vary from
culture to culture.

The capricious nature of cultural conceptions of development is
clearly indicated in the following illustration showing changes in
the ideal female figure during the past 100 years in the United States.

At the time of the Civil War, a small, tightly nipped-in waist, and
highly corseted figure strolling down the wooden sidewalks made the

1

boys whistle. By the 1890's, however, the ideal female shape accentuated certain essential physical features both fore and aft. Yet only thirty years later the accentuated figure was a low prestige item. The flapper of the 1920's was a rather sticklike, flat-chested, hipless, scrawny, linear female. Today, in contrast to these earlier periods, we admire a more natural physical type, hopefully endowed with a generous amount of the basic attributes that characterize the sex.

Americans, of course, are not the only people preoccupied with physical appearance and ideal development. Different cultures select different aspects of the human anatomy for elaboration and emphasis.

In the primitive African tribe of the Ubangi, large lips were a mark of beauty for young ladies. To achieve this trait, an incision was made in each lip at an early age and small wooden discs were inserted under the skin. As the lips were stretched, progressively larger discs

were used, and in time the young Ubangi woman acquired the desired lip size.

A completely different feature of physical beauty was emphasized by the Guanes Indians of South America. In this group, the aim was to achieve a flat head that bulged in the back and receded sharply in front. Systematic deformation of the skull was achieved through a molding process initiated in early infancy. During the first few weeks after birth, the individual bones of the head can be manipulated and the edges of adjacent bones may even be overlapped. The Guanes Indians were aware of the plasticity of the infant skull and exploited it for cultural advantage. The method used was not unlike this simple

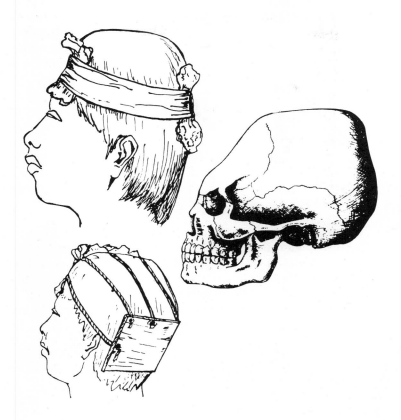

prescription for changing the shape of the head: The young infant is placed so that the back of his head presses against a flat board. A second board is positioned so that it presses against his forehead. The boards are bound together, with the binding tight enough to maintain a constant pressure on the skull but not so tight as to do injury to the brain.

As the child's skull grows, the head becomes systematically deformed and flattened. Progressive deformation results because the growing brain puts pressure on the inside of the skull and the skull must enlarge to accommodate it. Because of the pressure of the boards, normal enlargement of the skull at the forehead and the back is prevented; thus the enlarging brain forces compensating growth at the sides.

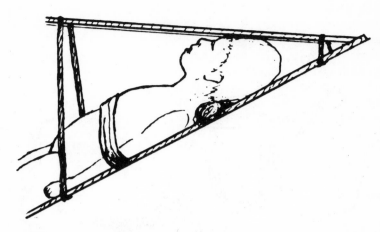

The Chinese also practiced systematic deformation of parts of the skeleton. A few generations ago, for example, the size of the feet in the female was a matter of prime concern. No Chinese gentleman of upper class would have married a woman with large feet. Although the Chinese could not actually stop the growth of the bones, they could distort the shape of the feet by binding them, thereby applying the same principle used by the Guanes Indians.

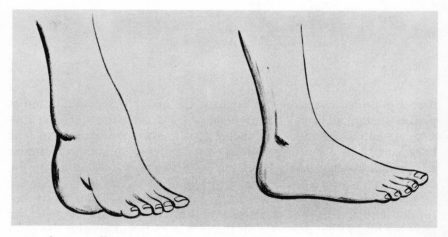

We have sufficiently illustrated the fact that every culture has its own idealized version of development and physical beauty. The fixations of other cultures may appear bizarre and strange to us, but they are rivaled by our own preoccupation with physical appearance. We clip, bob, and reshape our noses; we make slices in our skin and stretch it taut like a plastic bed sheet. If our ears stick out, we may pin them back against our heads. And there are other drastic surgical techniques too delicate to mention.

In each society relatively few are fortunate enough to be biologically endowed with physical traits that match the cultural ideal. Further, it is a fact that cultural gimmicks cannot fully succeed in recasting the individual into the ideal physical mode. Physical appearance is basically a function of biological development, a function of inherent biological processes that are operative from conception to death. In exploring these processes, we shall consider two key concepts: *growth* and *maturation*. These terms frequently are used interchangeably but they are not synonymous. Precisely what do they mean?

Conventionally, growth is regarded merely as an increase in size. But the conventional concept is not a sufficient characterization of growth. If it were, the infant would grow like an expanding balloon and the adult would simply be an enlargement of the infant. The accompanying illustration shows an actual mistake performed by nature in which growth was treated as simple size increase. The skull on the left is that of an infant; the one on the right is a young adult.

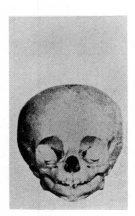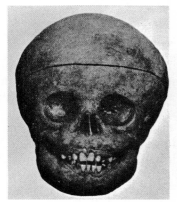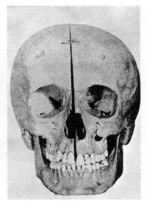

Although the skull in the middle is larger than that of the adult, it has retained the proportions of the infant. In this skull, the brain case is roughly seven times the area of the face, essentially the same ratio as that of the infant. In contrast, the adult ratio is roughly three to one, because of the disproportionately greater growth of the face.

Proportionate increase in size produces a grotesque distortion of normal growth. This is further illustrated by comparing the bodily

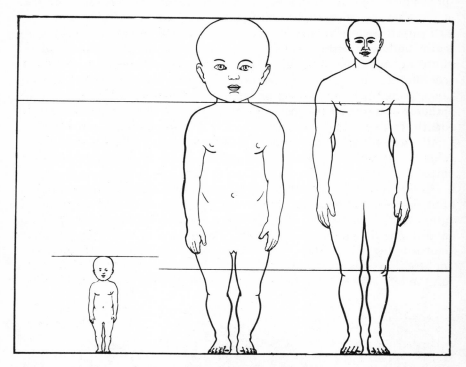

proportions of an infant and an adult. In the picture, the infant on the left represents the average length of a newborn male baby. If growth were only increase in size, the individual at adulthood would approximate the figure in the center. Obviously, this is not consistent with the proportions of a normal adult, as illustrated on the right.

From infancy to adulthood, the various parts of the body grow at different rates, resulting in proportional change. This phenomenon is called *differential growth*. The head of the infant, for example, constitutes one-fourth of total height but the head of the adult is only approximately one-eighth of adult stature. Consequently, while the head is actually growing larger, it undergoes a *relative decrease* in relation to total body size. In contrast to the head, the legs greatly increase in relative length, comprising approximately 35 percent of the total stature of the newborn infant and 50 percent of adult height. Although the trunk grows larger from infancy to adulthood, it remains essentially the same in relative size. During the whole period of growth each organ enlarges at its own particular rate. The difference in the growth rates of the various organs and structures is responsible for the attainment of normal adult proportions.

Thus far we have introduced two attributes of growth: *increase in size* and *differential growth*. To complete our definition we must add a third factor: growth can also be a *decrease in size*. As part of their normal growth, a number of structures and tissues of the body actually become smaller in size after adolescence. The thymus, parts of the spleen, the intestinal lymphoid masses—in fact, most of the organs composed largely of lymphoid tissue—evidence a *decrease* in size or volume after adolescence.

All three aspects of growth—size increase, size decrease, and differential growth—are illustrated in the accompanying graph. This graph depicts the curves of growth of four different tissues of the body. The amount of growth at each age is expressed as a percentage of the adult attainment at 20 years of age. Note the curve of growth

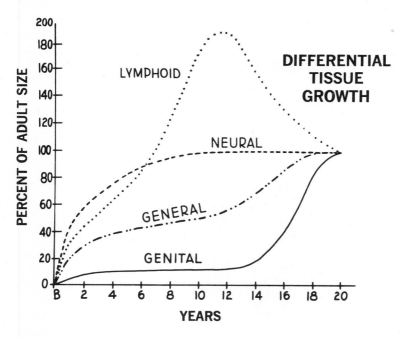

for lymphoid tissues. This curve reaches nearly 200 percent of adult size at 12 years of age and then undergoes a sharp decrease until it is reduced to 100 percent attainment at adulthood. In contrast to the lymphoid curve, which shows *size decrease*, the curves of growth for the neural tissue, the body generally, and the sex organs reflect markedly different rates of *size increase.*

The neural curve of growth, representing the brain, the spinal cord, and the eyeballs, rapidly achieves final size; in fact, the brain has achieved approximately 90 percent of its adult size by the time the child is 6 years old. The curve of growth for the body generally progresses much more slowly and does not reach 100 percent attainment until it approaches adulthood. The genital curve indicates that the primary and secondary sex organs undergo essentially no change in size from approximately 2 years of age until adolescence, at which time there is a rapid and dramatic increase leading to adult attainment.

All three aspects of the definition of growth refer either to dimensional or volumetric change in tissues, organs, and structures of the body. Growth, therefore, is a quantitative phenomenon and can be measured with a linear or a volumetric scale.

Maturation is the counterpart of growth. Sometimes this term is used to denote maturity or immaturity of social behavior, or the appropriate internalization of acceptable modes of behavior. For the biologist, however, maturation refers to the emergence of new tissues, new organs, and new structures, and to their unfolding in an orderly and predictable fashion during the life of the organism. It also refers to the new levels of physical functioning that the maturation of the tissues makes possible.

The picture below dramatizes the nature and significance of matura-

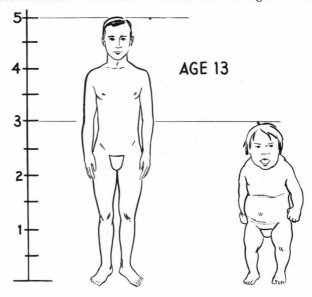

AGE 13

tion. It contrasts normal and retarded maturational patterns. Although both boys are 13 years of age, the boy on the right has experienced a drastic arrest in development—he has retained the appearance of a grotesque infant. Having suffered from a thyroid deficiency from birth, this boy is classified as an untreated cretin. When the thyroid hormone is not supplied in sufficient amounts by the thyroid gland or through hormone therapy, normal maturation is impaired.

In normal maturation, qualitative changes take place in all systems of the body. As a way of amplifying the definition of maturation, let us examine one of these systems, the skeleton. Have you ever thought about the number of bones in the human skeleton? The popular conception holds that there are 206. We would be closer to the truth if we turned these digits around and made the figure 602. Even then we would be underestimating the total number. During the life span of the organism, 806 discrete bone centers make their appearance. At different times from conception through old age, the human skeleton contains a changing number of bone centers. While 270 bones appear in the skeleton before birth, in the young child the number increases to 443. Then the trend is reversed. By young adulthood the number is reduced to 206 and continues to decrease until very old age, when there are fewer than 200 bones in the skeleton. Thus the number of bones varies according to the stage of maturation; while some bone centers are fusing together other bone centers are just appearing.

To illustrate the nature of skeletal maturation, we shall examine an x-ray picture of the hand of a newborn infant and compare it with that of a 6-year-old and that of a young adult. The hand of the newborn contains a simple arrangement of bones that are widely spaced,

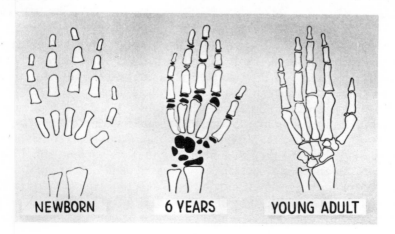

NEWBORN   6 YEARS   YOUNG ADULT

while the 6-year-old's hand presents a complex pattern as a result of the addition of many new bone centers. In fact, 28 new bone centers appear in the hand and wrist during the first 6 years after birth. In the ensuing years, many of the small bones in the palm and fingers fuse with adjacent long bones; consequently, the hand of the young adult has fewer bones than that of the 6-year-old. By actual count, there are 21 bones in the hand and wrist at birth, 49 bones at 6 years, and 31 bones in the young adult.

Maturational changes in the skeleton are further illustrated by examining the bones of the skull. In the newborn infant, four bones are clearly visible as shown in the picture. The large space between the bones is the soft spot or the anterior fontanelle. In the young adult, however, only three bones are present on the top of the skull. The

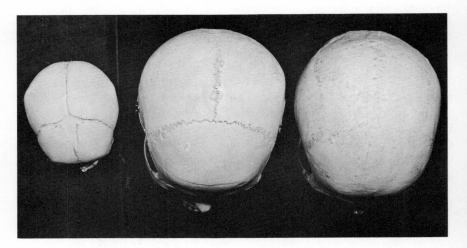

two at the front of the head have fused into one and the soft space has been filled in as a result of the growth of the adjacent bones. Actually, the fusion of the two frontal bones and the filling in of the fontanelle occurs between 12 and 18 months of age.

The picture on the right shows the top of the skull of a person in old age. During young adulthood, the bones forming the roof of the skull begin to fuse so that in old age the top of the skull consists of a single bone. Literally, then, we might say that we become boneheads; but more to the point, we have witnessed a series of qualitative changes, extending from infancy into old age, that we call maturation. It is evident that maturation of the skull is not a matter of size increase; rather, it involves a change in shape or configuration and the phenomenon of the fusion of bony centers. Maturation is a qualitative change not measurable by a yardstick.

Each maturational system of the body follows a predictable sequence of qualitative change. Bones, for example, do not appear in a random or erratic manner; rather, they follow each other in a fairly orderly fashion. Once an individual has reached a particular level of maturation, he cannot regress. Maturation is irreversible. A predictable sequence of maturational changes occurs in all human beings, irrespective of race or sex.

Since maturation is a qualitative phenomenon and is not measurable in a dimensional sense, status is assessed in terms of the age at which the individual achieves different maturity levels. Through the study of large populations of children and adults, standards have been established for assessing the rate of maturation. Each person is appraised in terms of how rapidly or how slowly he is approaching known maturational plateaus and whether he is an early or a late maturer.

In maturation, various arbitrary plateaus are known in advance, and their attainment can be assessed against time. In growth, the end point is not known in advance, but size increase can be measured with a dimensional scale.

We have stated that development refers to the totality of physical change from conception to death. Growth and maturation comprise the dual aspects of development. For clarification, we have considered these concepts separately, but as we study development in greater detail we shall see that these processes are bound together in an interesting pattern of interaction.

How we appear at any stage in our lifetime is a function of our development up to that point. The same universal processes are applicable to all human beings; they are predictable and demonstrable. Yet, as we view the products of development, the people around us, we are confronted by endless diversity in physical appearance. Cultural conceptions of appearance notwithstanding, each of us is distinct and different in his physical nature.

In order to understand the relationship between universal developmental processes on the one hand and the diversity of their expression on the other, we must examine the mechanisms of heredity responsible for individual variation on the central theme.

Chapter 2

A GAME OF CHANCE

When we observe a group of children of the same age in any society, we find that some are tall while others are short, some are heavy while others are thin, and some are physically mature while others are immature. Since the processes of development are universal, how are we to account for individual differences in every feature and attribute of the body? Are we pawns of a capricious nature or is there a sensible biological explanation behind the seemingly endless variation in physical appearance?

A number of factors have been advanced to account for individual differences in physical development. Particularly relevant in this respect are nutrition, climate, and disease. Although each of these variables influences the expression of the developmental pattern, they are not the primary determinants. Universally, the primary source of individual variation resides in the *hereditary mechanism*.

Integral to the reproductive process is a built-in mechanism that guarantees that each conception, each new individual, will be a unique creation. To understand this mechanism it is necessary to examine the nature and function of the egg and the sperm, because at conception the egg and the sperm contain the totality of hereditary biological endowment that the individual receives. At one time (during the Middle Ages) it was assumed that the sperm contained a homunculus, a perfectly formed miniature individual who needed only to be implanted in the egg and nourished to grow to the size of an infant at birth. We know today that this is not the case and that the real nature of the egg and the sperm can be understood by examining their size, structure, and functions. In size, the egg is significantly larger, having a diameter 25 times the length of the head of the sperm. In total volume, the egg is several thousand times greater than the sperm. Although minute, the egg can be detected by the naked eye, while the sperm can be seen only with the aid of a microscope. If the offspring reflected the relative size of the egg and the sperm, the

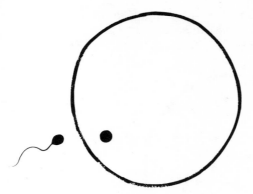

mother would play the predominant role in determining the nature of the child. However, the size discrepancy merely reflects the different functions each must perform.

The sperm may be conceived of as a stripped-down hot rod that contains the hereditary contribution of the father. It is mobile and agile so that it can convey the hereditary material to the egg. In contrast, the egg contains not only the hereditary contribution of the mother but also the yolk, which comprises the overwhelming volume

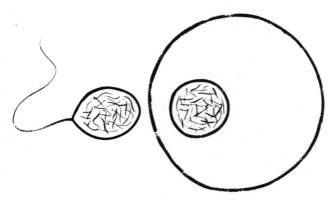

of the egg. The yolk provides the only source of nutrients for the developing embryo until the fertilized egg is established in the uterus and the food is furnished by the mother through the placenta.

The hereditary material contributed by the mother is localized in the nucleus of the egg, while that contributed by the father is contained in the nucleus in the head of the sperm; the nuclei are comparable in size and importance.

By enlarging the nuclei of the egg and sperm, which have been chemically stained, a number of short threadlike structures are revealed. If you count these structures in each of the nuclei, you will find that there are 23. Thus, the number of units in the nucleus of the

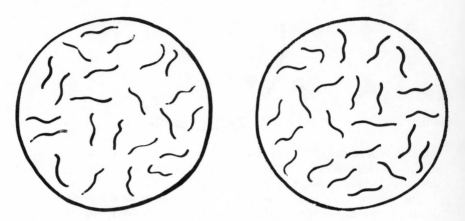

egg is matched by an equal number in the sperm, indicating that each parent makes a comparable hereditary contribution to each offspring.

These hereditary units are called *chromosomes*; and the name makes sense—colored bodies. Not many years ago, it was believed that there were 24 chromosomes in the nucleus of the egg and a like number in the sperm. Today, we know that there are but 23 in each. You might wonder how this mistake was made. The counting of chromosomes is much more difficult than appears to be the case from our illustrations. The chromosomes in the nucleus can be seen clearly at only one stage in the development of the cell. As a result of refinement of microscopic methods and techniques of staining, it is now possible to obtain clear images of the chromosomal material during this stage. Since the chromosomes in the egg differ in size and apparent shape, they can be identified and arranged in a standardized sequence from 1 through 23. This can also be done with the chromosomes in the sperm. After one sperm fertilizes the egg, the nuclei are combined. When the fertilized egg divides to create two cells, and during all subsequent cell divisions, each chromosome must make a duplicate of itself. This is necessary if each of the new cells is to receive a full set of 46 chromosomes. In this process chromosome number 1 from the egg pairs up with number 1 from the sperm. And right down the line all the chromosomes from the egg pair up with their partner chromosomes from the sperm. The illustration shows the part-

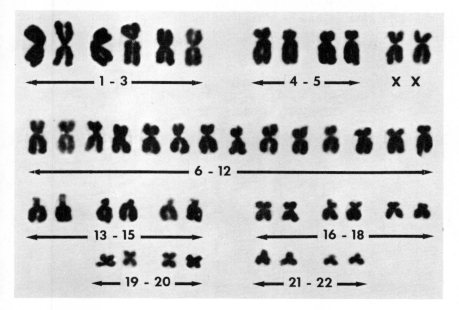

nership of the chromosomes preparatory to cell division. Each of the 46 chromosomes is in a state of duplication. This type of illustration is called a karyotype—a systematic arrangement of the chromosomes.

The normal human complement of chromosomes characterizing the fertilized egg and each and every cell of the body, with the exception of certain specialized cells such as the mature red blood cells with no cell nuclei, is 46.* However, if both the mother and the father passed 46 chromosomes on to their offspring, each new organism would receive a total of 92, twice the normal number; *their* offspring in turn would receive twice that number, or 184. Obviously this cannot be. Only 23, or one-half the total number of chromosomes we possess, may be passed on to our offspring by each parent. How is this problem of reduction solved?

---

*Infrequently, a child is born who possesses more than 46 chromosomes. This results in an unfavorable effect upon his development, appearance, and functioning. Many children diagnosed as mongoloids have been found to possess three rather than two chromosomes in the twenty-first pair. To date, no individual has been found to possess fewer than 46 chromosomes; presumably, embryos having an insufficient number of chromosomes are aborted at a very early developmental stage.

To demonstrate the solution, we must start with an *immature egg* that still contains the total complement of 46 chromosomes arranged in 23 pairs. One set had been transmitted to the girl from her mother and the other from her father. Both sets are present in every one of

her immature eggs. In contrast, each *mature egg* will contain but a single set of 23 chromosomes. It will be a mixed set consisting of some chromosomes from the girl's mother and some from her father. The selection of the chromosomes is a *game of chance*, and there are two rules by which this game must be played:

1. *Only 23 chromosomes may be selected from the 46 to make a mature egg or, correspondingly, a mature sperm.*
2. *One chromosome must be selected from every pair.*

This game of chance is nature's way of creating individual variation through the scrambling of hereditary materials. The biologist calls the process *reduction division,* an inexorable phenomenon that goes on in each and every one of us during the period of sexual maturity.

Let us actually play the game for a pair of prospective parents. The parents are labelled F for father and M for mother. We shall play the

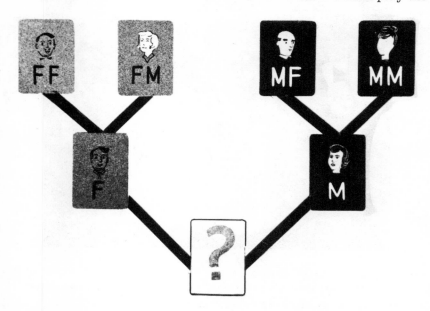

game first for the father. Each of his hundreds of thousands of immature sperm cells has 46 chromosomes, one set of 23 chromosomes from his father (FF) and the other set from his mother (FM). Our task is to create a single mature sperm in the father. Consider each set of chromosomes as a set of cards numbered from 1 through 23. In reducing from the 46 chromosomes to the 23, only 1 chromosome of each pair must be admitted. Whether chromosome number 1 comes from the father's father or from his mother is a pure chance phenomenon. The same holds true for all of the other 22 pairs. By shuffling the two sets of cards together, we provide for chance selection. From the deck of 46 shuffled cards, we deal out, face up, 23 cards. Each time a duplicate chromosome appears it is rejected, since only one from each pair can be admitted into the nucleus of the sperm. In the illustration, 8 cards or chromosomes bear the picture of the father's father while 15 are from his mother. Thus, the prospective father's mature sperm con-

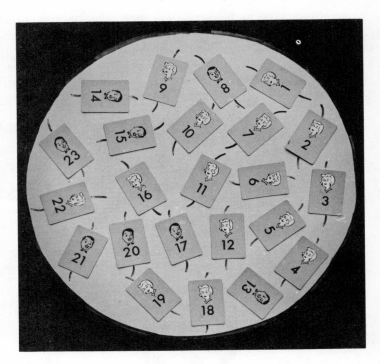

tains a combination of chromosomes that did not exist in either of his parents. If the two sets of cards were shuffled and the game replayed, the number of chromosomes representing each grandparent would almost certainly be different.

To create a mature egg for the prospective mother, the same procedure is followed. Through chance, 10 chromosomes from her father and 13 from her mother are selected.

What happens to the rejected partner of each pair of chromosomes from the immature egg and sperm? With respect to the sperm, the 23 chromosomes are not really rejected; they simply go into a partner sperm. The immature sperm divides into two mature sperms. Thus, 46 chromosomes are divided equally between the two mature sperms according to the rules of the game. In the case of the egg, however, the rejected chromosomes *are* discarded. These chromosomes enter another cell, which is called a polar body, a rudimentary egg that never matures and is eventually eliminated.

By playing the game of chance, a mature egg and a mature sperm are created. When the egg and sperm are joined, a new organism is conceived. This fertilized egg will contain 46 chromosomes, 23 from the mature sperm and an equal number from the mature egg. As our previous illustration shows, 8 chromosomes originated from the father's father and 15 from his mother; 10 chromosomes originated from the mother's father and 13 from her mother. The baby's development, then, will be controlled and directed by a new constellation of hereditary elements contributed by all four grandparents.

Is it possible for a mature sperm or egg to develop in which only one grandparent is represented? Although theoretically possible, this would be extremely rare. In fact, the likelihood of this occurring is equivalent to throwing 23 pennies into the air, with all of them falling heads up. The mathematical chance of getting all the chromosomes from the grandfather and none from the grandmother is only one in 8,388,608. The odds of a mature sperm fertilizing a mature egg with only one grandparent represented in each is fantastically small, only one in 70,368,744,177,664. The hereditary mechanism for reducing the number of chromosomes to a single set operates on a democratic principle, giving all four grandparents equal opportunity to be represented in their grandchildren.

To summarize, two basic processes are involved in the transmission of hereditary material from generation to generation: *Reduction division*, through chance selection, reduces the number of chromosomes to a single set of 23 in the mature sex cell. *Fertilization* restores the number of chromosomes to 46 and introduces an entirely new constellation of hereditary elements, one that has never existed before and almost certainly will never occur again.

Biochemists have made it possible to describe the nature and character of the chromosome and to determine how it works as a hereditary mechanism. The chromosome is regarded as consisting of a double spiral structure or double helix, as shown in the illustration. It is comparable to taking two strands of tape as they are released

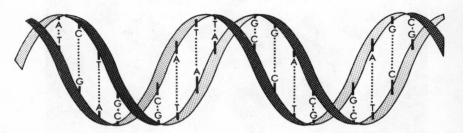

from the telegraphic ticker and twisting them together into a long spiral. The analogy with the ticker tape seems appropriate because both strands of the chromosome bear a continuous sequence of chemical codes. The alphabet in which the chemical code is written consists of four letters, or more properly, four chemical bases. The words expressing the message are all composed of three letters (three bases) drawn from this alphabet. The words and letters on one strand have a reciprocal relationship to their counterparts on the other strand, and together they comprise the definitive hereditary message. This structure, called DNA or deoxyribonucleic acid, is a substance unique to chromosomes.

The gene, regarded as the functional unit of heredity, actually comprises a segment of the continuous chromosome thread. This concept is reinforced by the fact that many biochemists believe that the chemical coding of the chromosome is interrupted periodically by punctuation. Some authorities are of the opinion that this punctuation marks the end of one gene and the beginning of the next. Each gene is believed to be a template providing the information necessary for the production of one of the thousands of proteins found in every cell. Thus the gene is a physical model, a segment of a giant complex molecule. It is a chemical organizer, a chemical catalyst, influencing, determining, and directing how the raw building blocks of the organism will become transformed into the final tissues, organs, and systems of the body. The genes themselves do not become transformed into the definitive organs and tissues; rather, they control chemicals that determine growth and maturation throughout the entire life of the individual. This amazing chemical apparatus is so minute that it cannot be seen even with the most powerful conventional microscopes known today. It is so small that all the genes needed to reproduce the present world's population of more than three billion people could be contained quite comfortably in the bottom of a teaspoon.

It has been estimated that there are roughly 80,000 *different* genes distributed among the 46 chromosomes of the fertilized human egg;* seemingly a small number to account for the millions of physical traits characterizing the human body. One might assume that a separate pair of genes existed (one gene from the mother and one from the father) for each and every physical trait and attribute. However, we know that most of our physical traits are the result of several pairs of genes working in concert. By teaming up with different constellations of genes, some gene pairs may participate in the development of a number of physical traits. Further, a single pair of genes may have several functions at different stages of the individual's development.

Each gene may exist in several forms or varieties, called *alleles*. The genes responsible for the development of many traits have but two alleles, while those responsible for other traits may have three or more. The gene for normal pigmentation of the hair, skin, and eyes, for example, has two alleles, a normal form and an abnormal form. When the abnormal allele is inherited from both parents, the result is an albino child. The gene for major blood type, on the other hand, exists in three varieties or alleles. The combination of alleles determines physical appearance and function.

As stated earlier, the hereditary material received at conception is transmitted in chromosomes obtained from the parents. Just as a child receives partner chromosomes from his parents, he also receives partner genes. Every gene on the first 22 pairs of chromosomes has a partner. The partner genes are located on the same pair of chromosomes and occupy the same position on the chromosomes. Working

---

*Although there are 80,000 *different* genes, the *total* number of genes is estimated to be approximately 800,000 because many if not all the genes are repeated many times.

in concert, they influence the particular aspects of development under their control.

Some of the genes on the twenty-third pair of chromosomes, the pair that which determines sex, do not always have partners. In the female, the chromosomes on the twenty-third pair, designated XX, are of equal length and the genes are always paired. But in the male, the twenty-third pair, consisting of an X and a Y chromosome, are of unequal length. The Y chromosome is shorter and lacks partners for some of the genes on the X chromosome. If the egg is fertilized by a sperm containing an X chromosome, then the offspring will be a female, since the twenty-third pair of chromosomes will be XX. If the egg is fertilized by a sperm bearing a Y chromosome, then a male will be conceived, as a result of the chromosome combination XY. It is interesting to note that the father determines the sex of the off-spring, since only the sperm can contain either an X or a Y chromosome. Although sperms containing either the X or the Y chromosome would appear to have equal chance to fertilize a particular egg, it is a fact that significantly more male than female babies are conceived. The reason why this is so is still unknown, but it would appear that the changing physiological state of the female during the critical period when conception can take place may be an important factor. In other words, the ovum may be more receptive to a sperm bearing a Y chromosome at a particular time.

Although most physical traits are the result of a number of pairs of

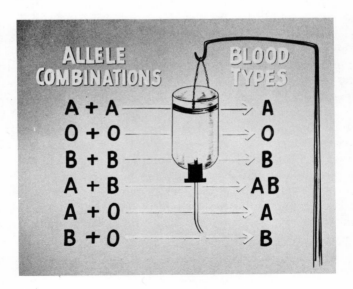

genes working in concert, some traits, such as major blood type, are determined by a single gene pair. The gene for major blood type exists in three varieties; therefore, this gene has three alleles, designated simply as A, B, and O. A number of hereditary combinations is possible. If an A allele from one parent is matched with an A allele from the other parent, the result is Blood Type A. An O allele matched with another O allele produces Blood Type O. Matching a B allele with another B allele results in Blood Type B. This condition of matching alleles is called *homozygous*, meaning a fertilized egg with the same alleles for a given gene pair.

If we match an A allele from one parent with an O allele from the other, we obtain Blood Type A because A is dominant over O. When matching B with O the result is Blood Type B since B is also dominant over O. In the final possibility, by combining an A allele with a B allele, we obtain Blood Type AB because these two alleles are equally dominant. When unlike alleles are combined, the condition is called *heterozygous*. In the heterozygous state, the results are the same regardless of which allele is transmitted by either parent; thus, the blood type is the same whether an A allele is received from the mother and an O allele from the father or the reverse.

From the family trees of our prospective parents it is possible to determine the blood type potentials of their offspring. The father inherited an O allele from his father and an A allele from his mother,

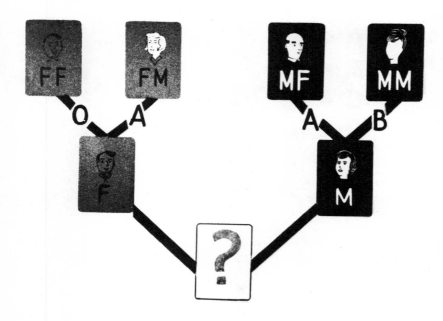

while the mother inherited an A allele from her father and a B allele from her mother. To each of his offspring, the father may pass on an O or an A allele—one or the other, but not both. Comparably, the mother may pass on either an A or a B allele.

Taking one allele from each pair of chromosomes from each parent, four combinations are possible, as shown in the accompanying illustration. These parents could have children with the following allele combinations and blood types: A with A forming Blood Type A; A with B, Blood Type AB; O with A, Blood Type A; and O with B, Blood Type B. Thus it is possible for our prospective parents to have three children each represented by a different blood type—or a dozen children with the same blood type. Although the parents have the potential for creating children with Blood Type A, the genetic makeup of these children might be different. One child could have allele combination AA (Blood Type A) and be homozygous, while another child could have allele combination OA (Blood Type A) and be heterozygous. The geneticist refers to an individual's allele combinations as his *genotype* and to the resultant physical traits as his *phenotype*. Therefore, two persons with the same phenotype, as in the illustration below may have different genotypes.

AO+AB   BLOOD TYPE

A + A	A
A + B	AB
O + A	A
O + B	B

Let us return to the infant in our example and look at the hereditary material that he received at conception. Geneticists have only partially determined on which chromosomes specific genes are located, with the exception of the sex-linked traits on the twenty-third pair. However, to continue our game, we shall *assume* that the alleles for major blood type are located on chromosome pair 2. By turning over the cards for this chromosome pair in the egg and sperm, we can see the allele labeled on the back of each card. Chromosome number 2 in the sperm bears the A allele while the partner chromosome in

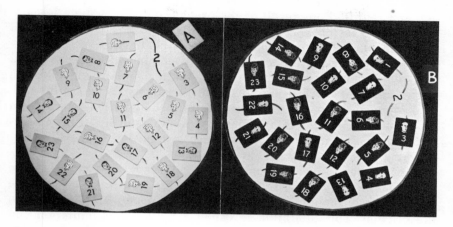

the egg contains the B allele. Thus the infant inherited the allele combination AB; his blood type is AB, the same as his mother's.

We have used but a single example, the alleles of the major blood groups, to illustrate the result of the chance selection of one partner chromosome rather than the other. Since each chromosome contains the alleles for more than a thousand different genes, theoretically it would be possible to repeat our experiment over a thousand times for each pair of chromosomes. It must be evident that each of the partner chromosomes would in all likelihood not contain the same allele forms for all of the thousand genes, but rather, each would represent a distinctive constellation of gene material.

Consider for a moment the possibilities of chance inheritance of at least 40 different blood factors and their alleles, which are capable of combining into 500 billion different blood genotypes. Then add all of the thousands of traits in our bodies, using all of the alleles in all of the chromosomes. The total number of possibilities of unique constellations of hereditary material is almost beyond human comprehension and belief. Yet as we have stated, it is the chance selection of one chromosome from each pair at reduction division and the subsequent recombination at fertilization which are primarily responsible for creating unending individual variation.

Because of the nature of the hereditary process, each individual begins life with a unique genetic blueprint. Once the blueprint has been established at conception, nature evidences a remarkable talent for duplication and repetition. The fertilized egg divides to create two cells; then these divide into four cells, and the four become eight. This process of *mitotic division* continues until the trillions of cells of the human body are formed. Each and every cell receives a true, exact, and faithful copy of *all* of the original hereditary material contained in the fertilized egg. In other words, each cell receives the full complement of 46 chromosomes represented in the fertilized egg.

Every time a cell is about to divide, beginning with the fertilized egg, the hereditary material behaves like a template and makes a mirror image of itself, using the chemical substances within the cell. One copy goes into each of the newly formed cells. This mechanism of duplication of hereditary material is fantastically efficient. Envision the task—all 800,000 genes must be copied exactly trillions of times!

But this incredible mechanism is not perfect; rare mistakes are made in the process of self-copying. When a mistake occurs, the chemical construction of the copied gene is somewhat altered from the original. This alteration is called a *mutation*. A mutation can occur at any stage in the developmental process, during the formation and development of the sex cells in the reproductive apparatus of the parent, or, after conception, during the formation of the body cells of the child. Once a mutation has occurred, all of the cells subsequently formed from that cell will, of course, contain the altered form of the gene. Nature is faithful even in duplicating her mistakes.

What is the significance of the mutant gene? Since the altered gene has a changed chemical structure, it induces an altered development and functioning of the tissues, organs, and structures under its influence and control. Gene mutations occur at a fairly constant rate in each species; a particular mutation will occur in man once in every 50,000 to 100,000 births. Most mutations are unfavorable. For example, in hemophilia or Bleeder's Disease, which can be caused by a mutation, the individual lacks the clotting substance in the blood. Color blindness, defects in the development of the skeleton and brain, deformity of organs and tissues, and a wide range of physical disorders have been identified by the geneticist as resulting from mutations.

Although the causes of spontaneous mutation are unknown, environmental or man-made conditions can speed up the normal mutation rate. A number of chemical substances as well as radiation from nuclear fallout or from X-rays are known to induce mutations. Consequently, the biologist often opposes the testing of atomic and hydrogen weapons in the atmosphere and recommends the cautious use of X-rays in medical diagnosis and treatment.

Not all mutations or new alleles are harmful or unfavorable. In fact, mutation is the way new alleles or variants of the gene come into being. This is the mechanism through which new traits are created and incorporated into the total gene pool of the population. The emergence of new alleles makes possible the further evolution of the species. The ideal, of course, would be to eliminate the harmful mutations and foster the beneficial ones.

Whether an allele is favorable or not depends on whether it will enhance or harm the development and survival of the organism in a *particular environment.* The way we develop and appear at any stage in our lifetime is always a function of the way the hereditary material is expressed through the medium of a particular environment. Even in the case of identical twins, the same genetic blueprint would evidence slightly varied expression in different environments. Since the hereditary material can be expressed only through the environment, it is clear that the environment plays a significant role in determining individual variation.

We are all similar as human beings because we hold a vast number of genes in common. We are all different because each of us has some constellation of alleles that is our own unique heritage. Thus the starting point in both our alikeness and our difference is to be found in the basic unit of heredity—the gene. On the one hand, the germ plasm transmits the established characteristics of the group; on the other hand, it provides for innovation and change through the unique recombination of hereditary materials and through the introduction of new traits, which are tested in the environment.

The individuals who result from the interaction of the hereditary material and the environment come and go, but the group lives on. From the time that life began nearly three billion years ago, there has been an unbroken stream of transmission of hereditary material from generation to generation. This is the deathless dimension of life!

Chapter 3

GROWTH IN STATURE

The process of development is like a motion picture. From conception through death, the development of the organism comprises an unbroken stream in which the individual moves without discontinuity from one stage to the next. In order to study the details of development, however, it is necessary to stop the motion picture and examine bodily changes as if they were depicted in a succession of still frames. The human developmentalist has many techniques for stopping the growth and maturation of different systems of the body, as it were, so that their pattern of development can be observed more effectively.

Measuring a child's body size at different chronological ages is a way of stopping the motion picture of growth. It enables us to contrast and compare a given child's body size with that of other children of the same chronological age. When we look at a child's height at a particular chronological age, for example, we are in effect examining a still taken from the motion picture.

In this chapter, we shall consider an attribute of growth that is of major interest in our society—growth in stature. Our cultural preoccupation with stature raises a number of questions: Is the child's height normal for his age? How tall will he be when he stops growing? Is there a way to increase height? These questions can be answered by a systematic study of statural growth. Let us pretend that we have no prior knowledge of this subject. How might we accumulate data that would give us a picture of growth in height?

Logically, we might begin with a newborn infant. We could go to a hospital and prevail upon the attending physician to permit us to measure the infant's length. During the first 18 months of life, stature necessarily is measurement of the *length* of the baby lying down. The most efficient method for determining length requires that the infant be placed on a measuring board equipped with sliding ends. We fit the instrument to the baby by bringing the headboard in to touch the

head and the footboard in to touch the soles of the feet. Reading from the scale, we observe that the infant is twenty inches long. We may now proceed with our study by measuring the child's height or length at equal time intervals. Each time a measurement is taken it is entered on the growth record until the child is able to stand erect. Then we abandon the horizontal measuring board and determine height through the use of a vertical measuring board, as shown in the accompanying illustration.

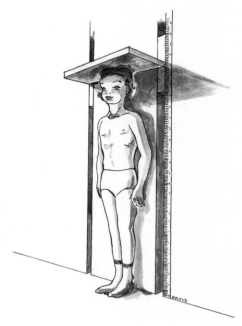

To obtain an accurate measurement of the child, he must stand against the board with his feet together, his knees not bent, his heels, buttocks, and shoulders touching against the back of the board, and his head oriented so that he is looking straight ahead. The horizontal board is then placed in firm contact with the top of his head. To ensure a precise measurement, we use sufficient pressure to compress the hair, while reading his height off the scale.

On the face of it, the technique for measuring stature seems absurdly simple; obtaining an accurate picture of a child's growth in stature, however, involves more than meets the eye. For example, if a child is measured after a hard day's play he will be appreciably shorter than he was in the morning, following a good night's sleep. And if the child is measured at very frequent time intervals, say a week or even a month apart, you may find that periodically he seems to be growing shorter rather than taller. Why does this erratic pattern occur?

During each day, our statures are subject to considerable fluctuation, depending on the nature of our activities. This fact has been demonstrated by an interesting set of experiments that you could repeat for yourself. In these experiments, children were measured early in the afternoon, following a rest period, before they could become fatigued from play. They were then induced to engage in vigorous activity and were subsequently remeasured. After play, the children were found to be appreciably shorter; after rest, they had recovered most of their lost stature.

These experiments offer dramatic evidence that fatigue is an important factor influencing stature and posture as well. All individuals experience *diurnal fluctuation*: our statures run down during the day and come up again after a night's rest. In the study cited, the average recovery after two hours of sleep was equivalent to 2 months' growth for children around 4 years of age.

These findings have important implications for measuring stature. To be accurate, all measurements must be made at the same time of day and under comparable conditions. An optimal period is a 3-month interval. When taken at extremely short time intervals, measurements will inevitably be influenced by extraneous factors, such as fatigue. Further, the inherent error in all measurement will obscure the small increment of growth that has occurred.

Having touched upon some of the variables influencing the measurement of stature, we now return to the task of determining the nature of statural growth from infancy onward. The table indicates the record of growth in stature for one young man from birth through

RECORD of GROWTH	
AGE YRS.	HEIGHT INCHES
BIRTH	2 0.0
1	3 0.1
2	3 4.6
3	3 7.9
4	4 0.5
5	4 3.3
6	4 5.6
7	4 8.0
8	5 0.6
9	5 2.5
10	5 4.5
11	5 6.3
12	5 8.2
13	6 0.5
14	6 2.9
15	6 5.0
16	6 7.0
17	6 8.0
18	6 8.5
19	6 8.7
20	6 9.0

20 years of age. His total growth in height during this period was 49 inches. In the first two years of life he grew more than 14 inches and achieved roughly half his adult stature; in the 20-year interval, his stature increased nearly three and one-half times.

Numerical data listed in tabular form are difficult to interpret; they do not easily convey the pattern of growth. To make the pattern apparent, we must convert the table into a graph by plotting the subject's height at each age. In the graph, chronological age is represented on the horizontal scale and height in inches on the vertical scale. The line formed by connecting all the height measurements is complexly curved, indicating that the rate of growth was not constant. If the individual had grown the *same* amount during each of his 20 years, and if the rate of growth *had* been constant, the line would be straight. The curved line indicates the periods when an individual is growing more rapidly and periods when he is growing less rapidly.

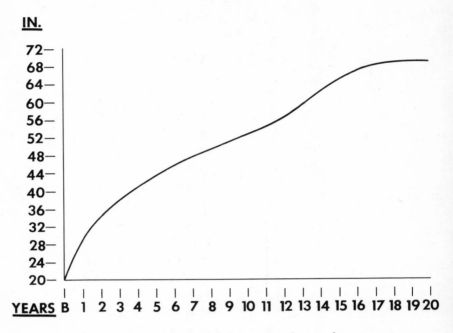

IN.

```
72—
68—
64—
60—
56—
52—
48—
44—
40—
36—
32—
28—
24—
20—
```

YEARS B 1 2 3 4 5 6 7 8 9 10 11 12 13 14 15 16 17 18 19 20

From birth to approximately 3 or 4 years of age, the curve goes up rather steeply, but nevertheless, it is humpbacked; that is, the child is growing rapidly but his growth is slowing down. From around 5 years of age to about 11 or 12 the curve is rather flat; growth at this time is fairly constant. The sharp upward turn at age 13 indicates that the rate of growth has accelerated and that the child is now growing more rapidly. In this adolescent phase, the individual is growing at successively faster rates during equal periods of time. Then the curve undergoes a radical change in direction and begins to flatten out. The child's growth is slowing down as he begins to approach his terminal stature at young adulthood.

In summary, the graph illustrates that the pattern of growth from birth to 20 years of age is characterized by three phases. After birth, a *decelerating phase* is evident and the curve is convex on top; growth is rapid but slowing down and becoming essentially linear. An *accelerating phase* occurs at adolescence when the curve is concave on top—growth is speeding up. Finally, another decelerating phase occurs when growth begins to slow down as young adulthood is approached. In the accompanying illustration, the phases of growth have been highlighted.

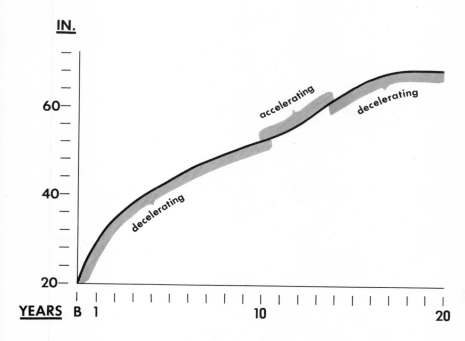

Earlier, we stated that development covers the total life history from conception to death. The period from birth to young adulthood is but a segment lifted out of the totality of development. In the graph, we have omitted the prenatal period, which represents the most dramatic phase in the development of the organism. In a mere time span of 9 months, tremendous size changes occur and the human organism grows from a microscopic embryo to an infant of approximately 20 inches in length.

The following illustration depicts the size of the embryo at different ages, in days and weeks, subsequent to conception, reflecting the dramatic increases that occur. The scale in the lower portion of the illustration, which represents the period beyond 90 days has been reduced, but the relative size differences have been maintained. The superimposed curves contrast and highlight the shape of the curve of growth characterizing each of the periods.

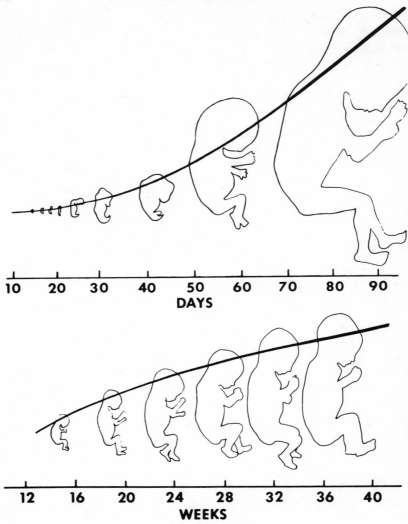

Since we cannot get inside the uterus and take actual life measurements of a child's prenatal growth, you might wonder how the size at each month is determined. Studies have been made of stillborn embryos and fetuses at different ages, and from these reports we have a good idea of the average length of the fetus at each month from conception to birth. The embryologist calls the developing organism an embryo during the first 2 months after conception and a fetus during the last 7 months.

Using the data drawn from many studies of embryos and fetuses, we have constructed a curve of growth representing the length of the organism during the period of intrauterine life. The curve shows an accelerating phase during the first 3 months of postconception life. In the last 6 months of intrauterine life, however, the curve reflects a decelerating phase despite the fact that growth is still extremely rapid. Thus, growth in length of the human embryo and fetus is

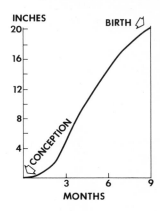

characterized by two phases, the first accelerating and the second decelerating.

In the next illustration, the prenatal curve of growth has been connected to the postnatal curve constructed earlier. One significant fact

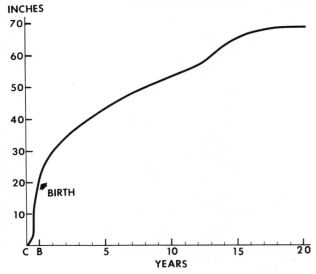

stands out in the composite curve: the decelerating phase of infancy is but an extension of the decelerating phase characterizing the last 6 months of fetal growth. Thus, it can be seen that birth is actually an arbitrary event in terms of the growth of the organism in length.

We still have not constructed a complete curve of growth. Does the person increase his stature after age 20? Using composite data, the curve may be extended through old age. From 20 to approximately 50 years of age stature remains fairly constant. After the age of 50,

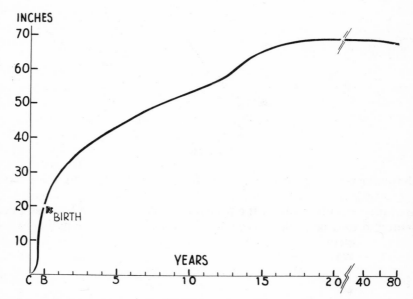

slowly and almost imperceptibly height begins to decrease, and this trend continues through old age. The little old man and the little old lady are not optical illusions. But is the statural change that we see in old age actually a decrease in size? Do the bones actually get smaller? No, the bones do not shrink; with increasing age the cartilage cushions between the vertebrae become compressed and lose their resiliency. Further, we lose muscle tone and the ability to hold ourselves erect. Our statures are shorter, but this decrease is not a true attribute of growth.

A curve of growth has been constructed from conception through very old age. We have seen that this curve consists of four phases, two accelerating and two decelerating, and that these phases appear in a regular order. The first accelerating phase and the beginning of the first decelerating phase occur in intrauterine life. The second accelerating phase begins in early adolescence and is followed, in turn, by the second decelerating phase leading to young adult stature. The decrease in stature evident in old age may not properly be regarded as a phase of growth since it is a function of posture.

Is this curve of growth representative of people generally? Does it have universal validity for people of all races and cultures? To answer these questions, appropriate samples from different populations have been studied and reliable methods of investigation have been formulated.

The human developmentalist has evolved two major methods for studying growth: the *longitudinal* and the *cross-sectional*. In the longitudinal approach, the same children are studied during the entire period of investigation. For example, a population of newborn infants may be studied at regular intervals until they reach the age of 20 or until they stop growing. At that time, the data would permit the investigator to construct precise curves of growth for each and every child in his study, as well as for the group as a whole. However, the lengthy time commitment is a major disadvantage of the longitudinal method. Investigators, therefore, often use the cross-sectional method, which permits the collection of data in an appreciably shorter time period.

In the cross-sectional method, a different group of children is selected to represent each chronological age. Children of different chronological ages may be measured concurrently and the data quickly gathered. However, this method also has its limitations, because it is assumed that children selected to represent each chronological age are comparable in terms of socioeconomic background

and genetic or racial origin. One would not attempt to construct a curve of growth based on a sample of 2-year-old Scandinavian children, 3-year-old Belgian Congo Pigmies, and 4-year-old Eskimos. The absurdity of lumping such diverse populations becomes readily apparent when the adult representatives of these three groups are compared.

Let us now examine a typical curve of growth based on a cross-sectional study conducted at the State University of Iowa. This curve represents the average height of 1,500 Caucasian boys at each age

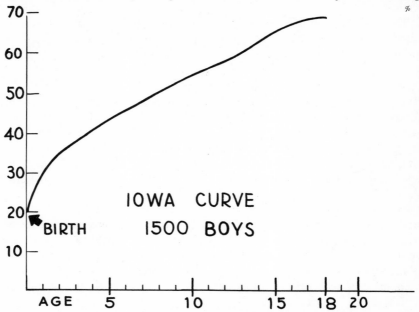

through 18 years of age. They have been drawn from a middle-class socioeconomic background. The pattern of the curve representing these boys evidences the same phases of accelerated and decelerated growth exhibited by the hypothetical individual we described earlier.

In order to determine whether girls also exhibit a similar pattern, we have superimposed an average curve of growth for Iowa girls on the curve for boys. At first glance, the two curves seem quite dissimilar. The girls' curve flattens out much earlier and more abruptly. The average girl reaches terminal stature several years earlier than the average boy. Further, during early adolescence the average girl is much taller. On the other hand, during young childhood and by the end of adolescence, the average boy is taller than the average girl.

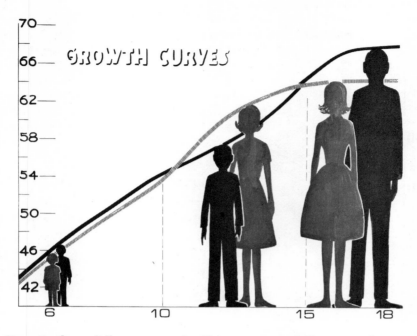

GROWTH CURVES

Despite these differences, a significant comparability exists between the two curves. The *form* of the curve is essentially the same for both sexes; both manifest the same phases of growth in the same sequence. Although the *timing* and *amount* of growth are different, the pattern is the same.

We have seen that there is a basic pattern of growth in stature that characterizes Caucasian children. But, is this pattern of growth characteristic of children of Negroid and Mongoloid races as well? In other words, is there a basically human curve of growth characteristic of the entire species? In order to answer this question, we shall compare curves of growth for Caucasian, Negro, and Mongoloid children.

Our comparison of these groups is based on data collected during the decades of the 1920's and 1930's from American Negro and Caucasian boys and from native-born Japanese boys. The three races differ in height at each age; the Caucasians are slightly taller than the American Negroes, while both of these groups are strikingly taller than the native-born Japanese. Comparable data collected in Michigan in the 1950's, on the other hand, show that the Negro males achieved a greater adult stature than the Caucasians. These differ-

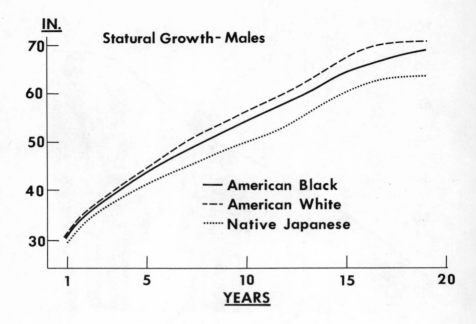

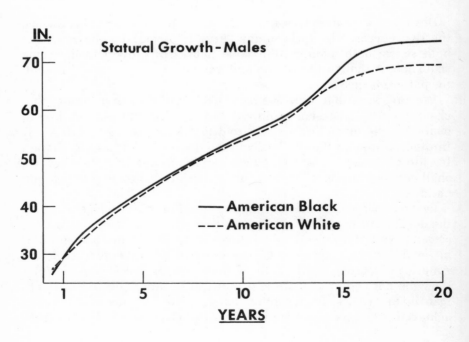

ences, however, do not mask the fundamental similarity in the pattern of the curves. From birth through 19 years of age, all three curves evidence the same sequence: an early decelerating phase followed by an accelerating phase at adolescence, which is succeeded, in turn, by a terminal decelerating phase leading to young adult stature. Thus, despite differences in the *magnitude* or amount of growth, the basic pattern of growth remains unchanged. Curves of growth of diverse populations reveal dramatic differences in size attainment, but when the size differences are stripped away, the same characteristic pattern of growth is evident in all races. There is a universal curve of growth in stature for the human species.

An average curve of growth summarizes a whole population in a single sweeping line, but when we look at specific individuals in these populations, a vast array of size differences at each chronological age is encountered.

This dilemma is easily resolved when it is noted that the *average* pertains to the population, not to the individual. When a point is plotted on a curve for 10-year-old Caucasian boys, it is but a single dot, a summation of all 10-year-olds. The average masks individual differences and reduces a population of children to a single common

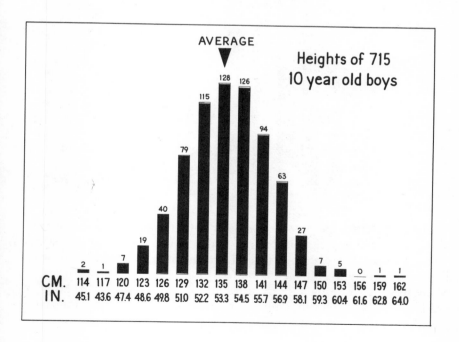

denominator. Since growth is an individual matter, in order to understand its nature, the mask imposed by the average must be removed
and the individuals who comprise the population revealed.

A bar graph is presented to show the distribution of stature for
10-year-old boys. These data are taken from a population of 715
boys reported in the Harvard growth study. Each bar represents the
number of boys at a particular height grouping. The two shortest
boys in the study are but 45 inches tall, while the tallest boy is 64
inches; thus the range in height in this population of healthy children is 19 inches. Although the average height of the boys is 53
inches, only 128 of them approximate this average.

By drawing a line connecting the tops of the bars, a bell-shaped
curve is created. This bell-shaped curve and the bar graph are ways
of depicting a *frequency distribution*; they show the frequency or

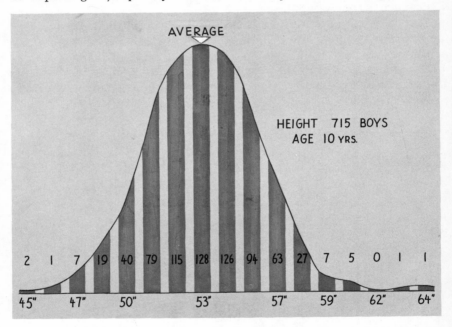

number of individuals present in each height grouping in the population. The curve is symmetrical and scales down evenly on either side
of the average. As we move away from the average, fewer and fewer
boys are represented at each height.

Let us now examine a distribution of heights in a group of 20-year-
old boys. This bar graph depicts a sample drawn from the U.S. Army
population at the end of World War II. It reveals the bell-shaped dis-

tribution characteristic of stature and the tremendous variation in heights observable in any age group.

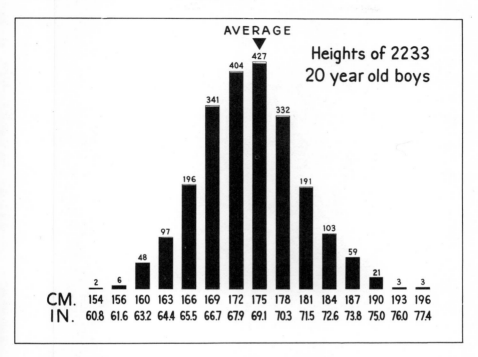

What are some of the factors that account for variation in statural attainment between and within human populations? Of the hereditary factors, the obvious one is sex. We have noted that boys and girls, while comparable in pattern, differ in magnitude of growth and in the timing of the phases of growth. Girls tend to be precocious, reaching the preadolescent spurt much earlier than boys and also achieving terminal stature at an earlier age. Sex is a source of variation that splits the human species in two in terms of the expression of the growth pattern.

Another hereditary factor responsible for population differences in statural growth is race. Some of the differences among various racial groups have already been illustrated. Within each race, it has been demonstrated that family lines manifest different rates of growth and varying potentialities for statural attainment. In some families, the members are consistently short, while in others they are consistently tall. Although the specific genetic factors are unknown, biologists hope some day to be able to isolate and identify the genes influencing growth in stature with the precision approximating that now possible in the identification of the different blood factors.

In an earlier chapter, it was stated that heredity never expresses itself in a vacuum, that the environment always mediates the expression of the genetic potential. A significant change in the environment, therefore, should exercise an influence on the expression of the growth potential. Nature and the physical anthropologists have conducted an interesting experiment that tests this contention. The three curves below represent boys of Japanese ancestry. The lower and middle curves refer to boys born and reared in Japan and measured

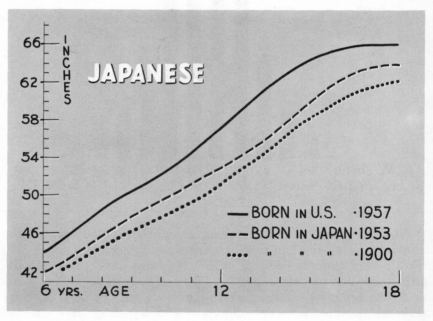

around 1900 and 1952, respectively. The upper curve represents boys born and reared in the United States and measured around 1957. In 1952 the native-born Japanese were systematically taller at all ages than their counterpart at the turn of the century. It is interesting that the American-born Japanese of 1957 were considerably taller than either of the native-born Japanese groups. The graph points up a size change, through time, in a single geographical area (Japan) as evidenced by a comparison of the two lower curves. It also points up a size change influenced by the environments of two different countries during the same time period, as shown in the middle and upper curves. Thus, the environmental differences between geographical areas, as well as environmental changes in the same geographical area over a period of time, are shown to influence the expression of the growth potential. This trend toward increasing body size, noted in many human populations, and presumably not the result of genetic modifications, is called *secular change*.

The enrichment of the diet is perhaps the most significant factor responsible for the increased growth in stature during the past half century in most of the civilized countries of the world. From the findings of research studies, we know that nutrition can have a pronounced and dramatic effect on growth in stature. For example, a few years ago in India, a group of investigators selected 46 girls from an orphanage as subjects in a research experiment on the effects of diet on growth. The girls were divided into two groups of 23 each and were matched for size so that each group contained girls of comparable statures. One group was labeled the *control group* and the other the *experimental group*. The average height of the control group at the beginning of the experiment was 45.83 inches, exceeding slightly the average height of 45.66 inches of the experimental group. In analyzing the diet available to the children in the orphanage, the researchers found it to be deficient in essential nutrients. For purposes of the experiment, the investigators permitted the control group to continue eating the standard diet of the orphanage, while the experimental group was given a supplemental diet containing the nutritional elements needed for healthy growth. Five months later the statures of the two groups were measured again. During the elapsed period both groups had grown, as one might expect. The accompanying chart shows that the average statures of the control group increased to 46.35 inches and the experimental group had increased to 46.62 inches. Thus, the experimental group, which was shorter at the beginning of the study, gained nearly one-half inch more stature than the control group.

· *Influence of Diet on Growth in Height*

	CONTROL GROUP (23 girls)	EXPERIMENTAL GROUP (23 girls)
Start of study	45.83"	45.66"
Five months later	46.35	46.62
The gain	.52	.96

Although a half-inch may not seem like a great amount, a statistical analysis of these data indicated that it is a highly significant improvement, unlikely to occur by chance more than one time in a hundred. This simple experiment shows clearly and conclusively that nutrition is an important factor influencing growth in stature. Messrs. V. Subrahmanyan, K. Joseph, T. R. Doraiswamy, M. Narayanarao, A. N. Sankaran, and M. Swaminathan who conducted the study, are to be commended.

A second factor responsible for secular increase in body size is control of disease. In diseases that have been studied extensively, such as diabetes, it has been found that when the disease is adequately controlled, a child will show essentially normal growth in height and weight, whereas when the illness is not controlled, growth is retarded. The tremendous developments in preventive medicine over the past several decades have made a significant contribution in freeing children from the suppressing effects of disease on growth.

A third factor influencing size attainment is climate. We do not as yet fully comprehend the mechanisms through which climatic factors control and influence body size. Physical anthropologists, studying body sizes of different human populations, have noted that there is a rough correlation between the body builds of various populations and the climates in which they live. People who live in arctic climates tend to be stocky and have relatively shorter appendages and relatively less surface or skin area; whereas those who live in tropical and hot desert areas tend to be thinner and have relatively longer appendages and more surface area relative to body mass. The squat, even globular appearance of the Eskimo is in sharp contrast to the tall, slim, elongated physique of the Nilotic (the Negro of the Sudan). These peoples live in radically different climates and temperature zones. The Eskimo must contend with intense arctic cold, while the desert Negro must face the implacable heat of the African sun. The Eskimo must preserve body heat, while the African Negro must dissipate it. In a sense both groups are confronted by a problem in geometry.

Where it is important to preserve body heat, it is advantageous for the individual to have a high ratio of body mass to skin area; where dissipation of body heat is essential, a low ratio of body mass to surface is advantageous. The reason for this is that heat is lost from the surface of the body. In geometrical terms, the three-dimensional figure having the smallest surface area to mass is the sphere. If a sphere were stretched out into the shape of a cylinder, the surface area in relation to the mass would be increased. Now, instead of a sphere and a cylinder, consider the globular Eskimo and the slim African Negro. Contrast and compare the geometric forms and the people in the accompanying illustration.

The effect of climate on body build in terms of the relationship of surface area to body mass has been studied in many mammals. Biologists have noted that warm-blooded animals that lived during the glacial periods, several hundred thousand years ago, were larger in body size than their present-day descendants. Observations of animals through time in different climatic areas resulted in the formulation of what has come to be called Bergmann's Rule and Allen's Rule.

Bergmann's Rule postulates that in cold climates the larger animal, like the larger sphere, will have a relatively smaller surface area and consequently less heat loss than the smaller animal. Allen's Rule amplifies the idea to show that reduction in the relative length of the limbs or the appendages affords still further protection against heat loss.

Our exploration of the secular factors shows that statural growth is highly susceptible to influence and that the environment dramatically affects the attainment of stature even in a generation or two.

To summarize, we have demonstrated that there is a pattern of stature that characterizes all human beings—a basic curve of growth that can be described from conception through old age. The characteristic curve of growth describes both males and females and peoples of all races and populations. At every chronological age there is considerable individual difference in size attainment, but if one ignores size increases as discrete entities and examines the manner in which size is achieved, it becomes clear that all normally developing individuals exhibit the same phases of accelerating and decelerating growth.

The statures of different human populations have not remained constant through the years, but rather, people have tended to grow progressively larger. This difference in stature over the centuries is not necessarily a function of change in the hereditary makeup of human beings but, to a large degree, reflects a more complete realization of the growth potential affected by environmental influences. Nutrition and control of disease are probably the two most significant environmental factors responsible for the dramatic increase in stature occurring in the last half century. Climate also influences physique and growth in stature.

We have defined stature as the vertical height of the body from the top of the head to the soles of the feet. But still the question may be asked, what is actually growing when we grow in stature? To answer this question, we must look within the body; we must examine the system that is chiefly responsible for the attainment of stature—the skeleton.

Chapter 4

THE LIVING SKELETON

The skeleton is a source of intrigue and mystery for young and old alike; it is a convenient symbol of Halloween and an almost indispensable prop in ghost stories. But in our present study we must eliminate all spooky considerations and examine the skeleton in its essential nature.

At first glance the skeleton appears to be a weird creation, a peculiar and curious contraption of oddly assorted parts—the product of a humorous inventor who worked with tongue in cheek. More careful perusal, however, shows that there is real order and purpose in its organization. The skeleton consists of many specialized parts that function in concert to meet the needs of the organism. It represents the culmination of 500 million years of evolutionary history and adaptation to different modes of life.

Almost exclusively, the skeleton determines the height of the person. When we say that we are so many inches tall, seldom do we realize that we have measured the height of our skeletons. Although height measurements treat the skeleton as if it were a single unit, actually stature is the result of the growth and maturation of a large number of bones. In order to understand how adult height is achieved, we shall discuss briefly each of the four major units of the skeleton that contribute to height.

The Cranium and the Face

Situated conveniently at the top of the skeleton is the skull, which consists of 28 bones in the young adult. The skull may be divided into two major parts: the cranium and the face. As the name implies, the cranium or braincase is a protective envelope that shields the brain against the physical buffeting and the blows encountered during childhood and the adult years. It is constructed to withstand pressure and percussion. In fact, we might compare the braincase with

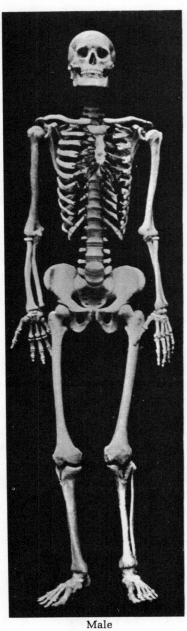 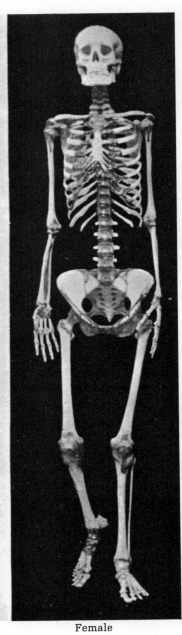

Male   Female

the Nautilus, the submarine in Jules Verne's story. Both are of double thickness, having an inner and an outer shell with bracing in between. The cranium has the further advantage of possessing a measure of resilience; the individual bones are capable of a slight degree of independent movement. On the inner surface of the cranium are delicate ridges and groovings representing the location of the blood vessels surrounding the brain. These markings show the extreme sensitivity of bone tissue to pressure. As the brain grows, the blood vessels leave their imprint on the inner surface of the cranium, as depicted in the accompanying illustration.

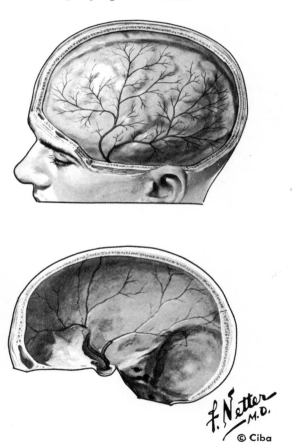

The face, the second component of the skull, houses and provides bony protection for the organs of vision and smell. It also contains a fresh air intake and an air exhaust, thus providing a built-in air-conditioning unit in the form of a nose. At the bottom of the face, the hinged jaw fosters two functions precious to all human beings: eating and talking.

The Vertebral Column

Below the skull we find the vertebral column or the backbone. The vertebral column consists of 32 spools, one on top of the other, held in position by ligaments and muscles. It is a fantastic arrangement, reminiscent of a child's tower of precariously balanced blocks. Yet the vertebral column is also like a flexible steel spring, holding the torso erect and permitting it to bend frontward, backward, and sideward, and even to rotate on itself.

The vertical arrangement of the vertebral column was not part of the original design of animals with backbones. Originally, these animals were aquatic forms; later, when they emerged on the land, they walked on all fours, and the vertebral column was essentially horizontal. The internal organs were suspended from the horizontal vertebral column like wash hanging on a clothes line. During the course of human evolution, man assumed an upright posture, and the vertebral column became realigned in the vertical plane. As a consequence, our internal organs project from the vertebral column like flags from a flagpole in a high wind. Man's upright posture has imposed a tremendous strain on his back. The internal organs tend

to slump down under the pull of gravity and must be held in place by ligaments. No wonder the popular exclamation, "Oh, my aching back!"

Many women in an advanced stage of pregnancy experience considerable back strain imposed by the extra weight of the developing fetus. The vertical arrangement of the back is not a good engineering design, but we are stuck with it. Surgical supply houses capitalize on this limitation and do a thriving business in maternity corsets and orthopedic braces. Although upright posture is an ingenious adaptation, nature has not fully solved the problem of reorganizing our skeletons to support the weight of the internal organs. We would have avoided many back problems if we had remained four-footed.

The Pelvis

The next unit of the skeleton is the pelvis, or the hip bones. The pelvis is attached to the base of the vertebral column and may be likened to a pan holding the internal organs and preventing them from dropping out of the bottom of the torso. Doctor W. M. Krogman has compared the pelvis to Grand Central Station because it must not only support the weight of the upper part of the body, but it must also serve as a channel for the digestive and reproductive systems and as an anchor for the muscles of the legs.

The female pelvis is constructed to permit the passage of the infant at birth. The woman with an unusually small pelvis faces a problem when she gives birth to an infant of normal size. It is the same problem that confronts the ambitious do-it-yourselfer who has arduously constructed a large powerboat in the basement of his modest home and then finds that he is unable to get the boat out of the house without performing a Caesarian operation. The woman with a small pelvis must also resort to opening up a wall. Fortunately, nature has taken

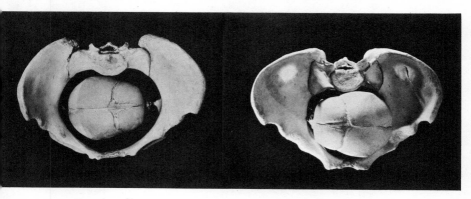

Female Pelvis   Male Pelvis

cognizance of the function of child bearing and has designed the
female pelvis so that the infant in most instances can pass through
the inlet or opening at birth without difficulty.

The inlet in the typical male pelvis, on the other hand, is so con-
stricted that it would not permit the passage of a newborn infant. An
interesting little experiment can demonstrate this fact. Take the
skull of a newborn infant and two pelves of comparable total size,
one male and one female. Try to pass the infant skull through the
two pelves. You will find that the skull will pass through the female
pelvis with ease. But when you attempt to pass it through the
male pelvis it gets hung up in the inlet, as shown in the illustration.
From this experiment, we see that total size of the pelvis is not the
fundamental consideration. It is the size of the inlet that is important.

The series of illustrations below depict the fundamental differences
in the shape and proportions of the male and female inlets. The total
width of the pelvis has been enlarged to the same size in both sexes.
The first picture shows the front view of the pelvis in each sex. In
the male, the pubic bones converge to form a narrow or acute angle;
the bones are pressed down, constricting the inlet. In contrast, in the
female the pubic bones flare out and form a wide or obtuse angle,
opening up the inlet.

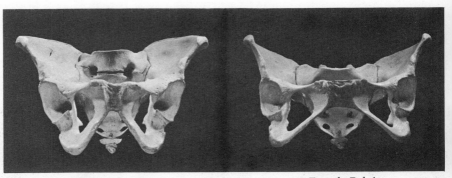

Male Pelvis   Female Pelvis

A second important factor contributing to the size of the inlet is
illustrated in the side views of the pelves. The sciatic notch in the
male is very narrow, and the end of the vertebral column is bent forward. In the female, on the other hand, the notch is extremely wide,

and the end of the vertebral column is arched backward. Thus, the female inlet has greater depth from front to back.

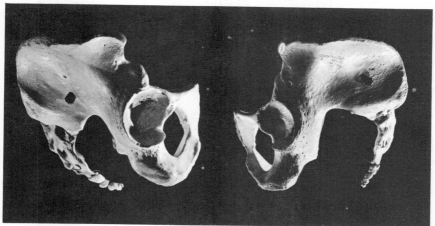

Male Pelvis   Female Pelvis

Examining the back view of the pelvis, we see that the central portion consists of five vertebrae fused into a single unit called the *sacrum*. This fusion of the vertebrae is nature's way of providing the pelvis with greater stability. The female sacrum is much wider than the male; consequently, the hip bones in the female are spread further apart, thereby widening the inlet.

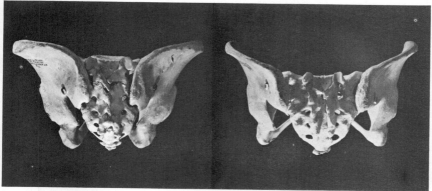

Male Pelvis   Female Pelvis

Thus, the three major factors that influence the shape of the pelvis are the angle of the pubic bones, the width of the sciatic notch, and the width of the sacrum. As a result of these factors, the female inlet is large and oval, while that of the male is small and heartshaped.

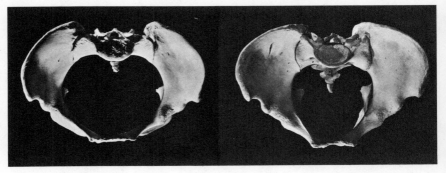

Female Pelvis   Male Pelvis

You may have seen in an issue of the Sunday supplement that the man of the future is depicted with an extremely large head and a small body. This assumes that human brain size will continue to increase as dramatically in the next million years as it has in the last million. Consider the implication: If the man of the future is to have a tremendous head, the woman of the future must have a tremendous pelvis, because the huge head of the infant must pass through it. The prospect of bearing bigger-brained children would be met with less than acclaim by the American female.

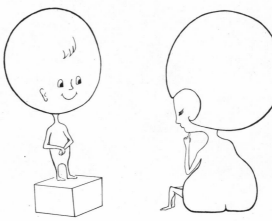

If you are an observer of anatomical differences, you have noticed that the female wiggles while she walks but the male does not. This trait is often assumed to be a cultural affectation, but actually the wiggle has an anatomical basis. In the female, the thigh bone is positioned more to the front of the pelvis than in the male. Therefore, in order to take a forward stride, the female must rotate her pelvis from side to side.

The Lower Limbs

Let us now examine the terminal unit of the skeleton, the lower limbs. The thigh is attached to the pelvis by a universal joint, which facilitates maximum mobility in all directions. The lower extremity, including the feet, consists of 30 bones, but only 4 of these contribute directly to vertical height. These bones are labeled as follows: the thigh or femur, the shin or tibia, and the talus and calcaneous, the two bones at the rear of the foot.

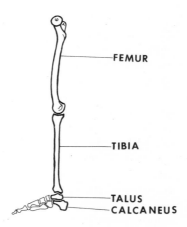

At your next visit to a zoo watch the feet of the gorilla or the chimpanzee and notice how effectively they are used for grasping. In man, the lower extremity has become specialized for bipedal or two-footed locomotion. The human foot is so committed to the task of carrying us about that it lacks the capacity for grasping objects. The large toe, which is highly mobile in the gorilla, has been tied to the adjacent toe in man by powerful ligaments.

In the discussion of the fundamental units of the skeleton, the arms and the rib cage have been omitted simply because they do not contribute to stature.

The Nature of Skeletal Development

Although the major units of the skeleton contributing to stature have been presented, we have not as yet explained *how* the skeleton develops. Lest the reader become frightened, we do not intend to pursue the formation of the skeleton bone by bone. We shall, however, in a general way, examine the nature of skeletal development. You may remember that by development we mean all of the normal predictable changes that occur in the body. To understand the nature of these changes in the skeleton, we must examine the mechanisms through which the bones grow in size and the sequence of qualitative changes through which the skeleton matures.

Maturation

At approximately 5 to 6 weeks after conception, the first bones of the body make their appearance, and with this event the skeleton is launched on a maturational career that is not completed until very old age and is terminated only at death.

The following maturational events occur during the development of the skeleton:

1. During the life career of each individual, some 806 bony centers make their appearance. These bones appear in a fairly orderly and predictable sequence, like actors in the wings following the cues of an unseen stage manager.
2. After its initial appearance, each bony center undergoes an intricate series of changes in form or shape. Like the precise unfolding of a classical ballet, these changes also occur in an orderly and predictable sequence.
3. Groups of bony centers fuse together at different times during the development of the skeleton. Again, this series of events takes place in an orderly and predictable way.
4. Before, during, and even after the bony centers have fused, the tissues of which the bones are composed undergo changes in texture and chemical composition.

The orderly sequence of qualitative change is the very heart of every maturational system. *All the maturational changes that occur in the skeleton are known in advance.* These changes manifest themselves in all normally developing individuals, regardless of sex or race. Nevertheless, individual variation is expressed in the rate or speed with which the person undergoes the series of known changes. Some individuals mature more rapidly in their skeletal development than others; each person exhibits his own rate of maturation.

Appearance of the Bony Centers

The first event in the maturation of the skeleton is the appearance of the bony centers. We have mentioned that during the total development of the skeleton 806 of these centers make their appearance. The first of these, the jawbones and the collarbones, are evident as early as 6 weeks after conception, while the last, such as the ridge on top of the hipbone, do not appear until puberty.

You may have noticed that at times the term bone and at other times the term bony center has been used. Actually, bones are formed from one or several bony centers. The bony center is the developing unit, while the bone is the finished structural unit.

Each bony center has its own developmental career characterized by a series of maturational form changes. The initial step in this career is called the *onset of ossification*, the time when the hard bony

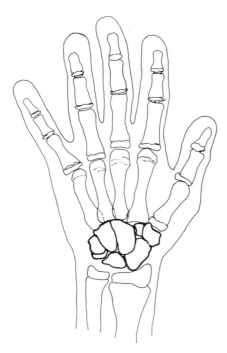

tissue is first laid down in the bony center. Following its first appearance as a tiny speck, each bony center grows in size and undergoes a series of transformations in shape. The change in shape occurs, in part, because the bones contact each other like pieces of a jigsaw puzzle.

Fusion of Bony Centers

The next major event in the maturation of the skeleton is *the fusion of bony centers* with one another. Fusion does not occur at the same time throughout the skeleton. Some centers fuse before others even appear. In fact, while some of the centers in the base of the skull begin to fuse during prenatal life, other centers in the brain case do not fuse until 70 years of age or more. Some bony centers preserve their separate identity throughout life and never fuse with adjacent bones. The small round bones in the wrist and ankle are excellent illustrations of staunch individualism.

Lipping

Fusion of the bony centers is not the end point in the maturational process. Beginning at about 35 or 40 years of age, the vertebrae in the back and the long bones in the arms and legs develop ridges or lips at their edges. This "lipping" is a natural arthritic maturational event, a normal aging phenomenon. In old age, the arthritic lipping of

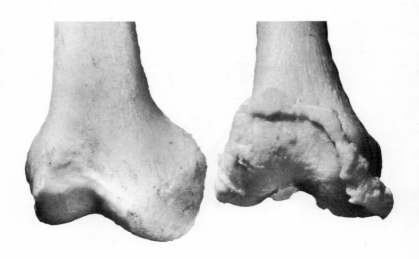

the long bones sometimes become extreme, and in pathological instances, the overgrowth of bone at the joints may even destroy the joint and weld adjacent bones together into one continuous block of bone. This also may happen in the vertebral column, where a number of vertebrae become united into a single structure, thus destroying the mobility of the back and preventing the individual from performing the graceful movements of which the normal human back is capable.

Changes in Composition of Bone Tissues

Another important maturational process in the development of the skeleton involves the progressive change in the *composition* of the bone tissue. When the bony centers are first formed, the tissue is coarse and irregular. During very young childhood, this primary bone tissue is removed and replaced by secondary bone tissue, which is more highly organized and uniform in texture. In young adulthood the bone tissue is dense, smooth, and homogeneous. Then, as the skeleton matures during middle and old age, the blood supply to the bones is progressively reduced, causing the bone tissue to lose its resiliency and become brittle and porous. The mineral elements of calcium and phosphorous are withdrawn from the bones, so that in very old people the flat bones are porous and translucent. Portions of them, such as the shoulder blade, are often paper-thin. For these reasons, the healing of a broken bone is a slow and laborious process in old age.

Timing of Maturational Events

We have discussed the major types of changes that occur during the maturation of the skeleton. These include the sequence and time of appearance of all the bony centers in the body; the progressive change in shape of the centers from the time they are first visible until they have undergone the modifications characteristic of very old age; the sequence and degree of fusion of the different bony centers; and finally, the changes in the composition and texture of the bones. Now let us examine these events in more detail, using as illustrations the six areas of the body that have been studied most extensively—the shoulder, the elbow, the hand and wrist, the hip, the knee, and the foot. Our knowledge of these areas is derived chiefly from the analysis of x-ray pictures taken of large groups of children at annual or semiannual intervals from birth through 20 years of age.

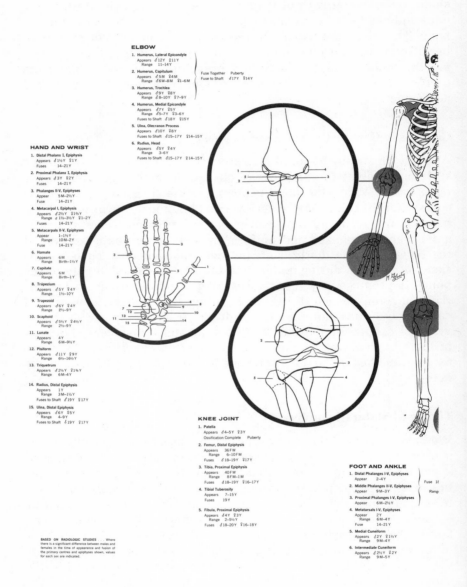

ELBOW

1. Humerus, Lateral Epicondyle
   Appears   ♂12Y ♀11Y
   Range   11–14Y

2. Humerus, Capitulum
   Appears   ♂5M ♀4M   Fuse Together   Puberty
   Range   ♂6W–8M ♀1–6M   Fuse to Shaft   ♂17Y ♀14Y

3. Humerus, Trochlea
   Appears   ♂9Y ♀8Y
   Range   ♂8–10Y ♀7–9Y

4. Humerus, Medial Epicondyle
   Appears   ♂7Y ♀5Y
   Range   ♂5–7Y ♀3–6Y
   Fuses to Shaft   ♂18Y ♀15Y

5. Ulna, Olecranon Process
   Appears   ♂10Y ♀8Y
   Fuses to Shaft   ♂15–17Y ♀14–15Y

6. Radius, Head
   Appears   ♂5Y ♀4Y
   Range   3–6Y
   Fuses to Shaft   ♂15–17Y ♀14–15Y

HAND AND WRIST

1. Distal Phalanx I, Epiphysis
   Appears   ♂1½Y ♀1Y
   Fuses   14–21Y

2. Proximal Phalanx I, Epiphysis
   Appears   ♂3Y ♀2Y
   Fuses   14–21Y

3. Phalanges II–V, Epiphyses
   Appear   5M–2½Y
   Fuse   14–21Y

4. Metacarpal I, Epiphysis
   Appears   ♂2½Y ♀1¾Y
   Range   ♂1½–3½Y ♀1–2Y
   Fuses   14–21Y

5. Metacarpals II–V, Epiphyses
   Appear   1–1½Y
   Range   10M–2Y
   Fuse   14–21Y

6. Hamate
   Appears   6M
   Range   Birth–1½Y

7. Capitate
   Appears   6M
   Range   Birth–1Y

8. Trapezium
   Appears   ♂5Y ♀4Y
   Range   1½–10Y

9. Trapezoid
   Appears   ♂6Y ♀4Y
   Range   2½–9Y

10. Scaphoid
    Appears   ♂5½Y ♀4½Y
    Range   2½–9Y

11. Lunate
    Appears   4Y
    Range   6M–9½Y

12. Pisiform
    Appears   ♂11Y ♀9Y
    Range   6½–16½Y

13. Triquetrum
    Appears   ♂2½Y ♀1¾Y
    Range   6M–4Y

14. Radius, Distal Epiphysis
    Appears   1Y
    Range   3M–1½Y
    Fuses to Shaft   ♂19Y ♀17Y

15. Ulna, Distal Epiphysis
    Appears   ♂6Y ♀5Y
    Range   4–9Y
    Fuses to Shaft   ♂19Y ♀17Y

KNEE JOINT

1. Patella
   Appears   ♂4–5Y ♀3Y
   Ossification Complete   Puberty

2. Femur, Distal Epiphysis
   Appears   36 F W
   Range   6–10 F M
   Fuses   ♂18–19Y ♀17Y

3. Tibia, Proximal Epiphysis
   Appears   40 F W
   Range   8 F M–1 M
   Fuses   ♂18–19Y ♀16–17Y

4. Tibial Tuberosity
   Appears   7–15Y
   Fuses   19Y

5. Fibula, Proximal Epiphysis
   Appears   ♂4Y ♀3Y
   Range   2–5½Y
   Fuses   ♂18–20Y ♀16–18Y

FOOT AND ANKLE

1. Distal Phalanges I–V, Epiphyses
   Appear   2–4Y   Fuse 1
2. Middle Phalanges II–V, Epiphyses
   Appear   9M–3Y   Rang
3. Proximal Phalanges I–V, Epiphyses
   Appear   6M–2½Y
4. Metatarsals I–V, Epiphyses
   Appear   2Y
   Range   6M–4Y
   Fuse   14–21Y
5. Medial Cuneiform
   Appears   ♂2Y ♀1½Y
   Range   9M–4Y
6. Intermediate Cuneiform
   Appears   ♂2½Y ♀2Y
   Range   9M–5Y

BASED ON RADIOLOGIC STUDIES . . . Where
there is a significant difference between males and
females in the time of appearance and fusion of
the primary centres and epiphyses shown, values
for each sex are indicated.

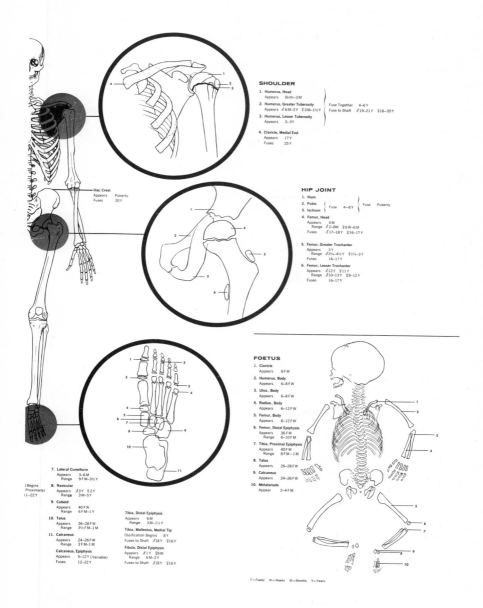

SHOULDER

1. **Humerus, Head**
   Appears   Birth–3 M

2. **Humerus, Greater Tuberosity**
   Appears   ♂6 M–2 Y ♀3 M–1½ Y   Fuse Together   4–6 Y

3. **Humerus, Lesser Tuberosity**
   Appears   3–5 Y   Fuse to Shaft   ♂19–21 Y ♀18–20 Y

4. **Clavicle, Medial End**
   Appears   17 Y
   Fuses   25 Y

HIP JOINT

1. **Ilium**
2. **Pubis**   Fuse   4–8 Y   Fuse   Puberty
3. **Ischium**

4. **Femur, Head**
   Appears   4 M
   Range   ♂2–8 M ♀6 W–6 M
   Fuses   17–18 Y ♀16–17 Y

5. **Femur, Greater Trochanter**
   Appears   3 Y
   Range   ♂2¾–4½ Y ♀1½–3 Y
   Fuses   16–17 Y

6. **Femur, Lesser Trochanter**
   Appears   ♂12 Y ♀11 Y
   Range   ♂10–13 Y ♀9–12 Y
   Fuses   16–17 Y

FOETUS

1. **Clavicle**
   Appears   6 F W

2. **Humerus, Body**
   Appears   6–8 F W

3. **Ulna, Body**
   Appears   6–8 F W

4. **Radius, Body**
   Appears   6–12 F W

5. **Femur, Body**
   Appears   6–12 F W

6. **Femur, Distal Epiphysis**
   Appears   36 F W
   Range   6–10 F M

7. **Tibia, Proximal Epiphysis**
   Appears   40 F W
   Range   8 F M–1 M

8. **Talus**
   Appears   26–28 F W

9. **Calcaneus**
   Appears   24–26 F W

10. **Metatarsals**
    Appear   2–4 F M

Iliac Crest
Appears   Puberty
Fuses   20 Y

7. **Lateral Cuneiform**
   Appears   3–6 M
   Range   9 F M–3½ Y

8. **Navicular**
   Appears   ♂3 Y ♀2 Y
   Range   3 M–5 Y

(Begins Proximally)
(1–22 Y

9. **Cuboid**
   Appears   40 F W
   Range   6 F M–1 Y

10. **Talus**
    Appears   26–28 F W
    Range   3½ F M–1 M

11. **Calcaneus**
    Appears   24–26 F W
    Range   3 F M–1 M

Calcaneus, Epiphysis
Appears   5–12 Y (Variable)
Fuses   12–22 Y

Tibia, Distal Epiphysis
Appears   6 M
Range   3 M–1½ Y

Tibia, Malleolus, Medial Tip
Ossification Begins   8 Y
Fuses to Shaft   ♂18 Y ♀16 Y

Fibula, Distal Epiphysis
Appears   ♂1 Y ♀9 M
Range   6 M–2 Y
Fuses to Shaft   ♂18 Y ♀16 Y

F=Foetal   W=Weeks   M=Months   Y=Years

Using the chart devised by Dr. R. Hugo Mackay, we have reproduced the skeleton with an enlargement of the six areas. Each of the bony centers is numbered so that it can be identified easily. By checking the appropriate column, the individual bone can be located and the time of its appearance and its fusion with an adjacent center can be determined. Sex is indicated by the appropriate symbol: ♂ for the male and ♀ for the female.

At first glance the illustration may seem a hopeless jumble, but taken a piece at a time, it can be easily interpreted. Examine the circle representing an enlargement of the hip joint. In this circle, numbers 1, 2, and 3, respectively, refer to the ilium, the pubis, and

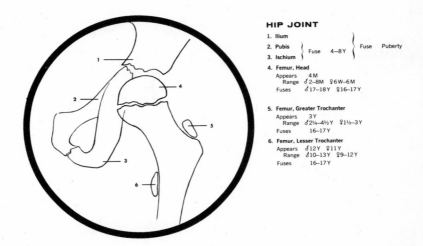

HIP JOINT

1. Ilium
2. Pubis   } Fuse 4–8Y   } Fuse   Puberty
3. Ischium }

4. Femur, Head
   Appears   4 M
   Range   ♂ 2–8M ♀ 6W–6M
   Fuses   ♂ 17–18Y ♀ 16–17Y

5. Femur, Greater Trochanter
   Appears   3 Y
   Range   ♂ 2¼–4½ Y ♀ 1½–3 Y
   Fuses   16–17 Y

6. Femur, Lesser Trochanter
   Appears ♂ 12Y ♀ 11Y
   Range   ♂ 10–13Y ♀ 9–12Y
   Fuses   16–17 Y

the ischium, the three bones that form one side of the pelvis. These bones ultimately fuse into a single unit. The fusion follows an orderly sequence: the pubis and ischium fuse together first; then these two fuse with the ilium. Now look at the elements at the upper end of the thighbone where the thigh connects with the hip. The long central portion forms the shaft of the thighbone; surrounding the shaft are three parts, numbered 4, 5, and 6. The ages at which these bony centers first appear and the time at which they fuse to the shaft are indicated in the diagram.

The shaft of the thighbone is not in the circle because it appears prenatally. The diagram of the fetus indicates that the thighbone or femur, labeled number 5, appears between 6 and 12 weeks after conception.

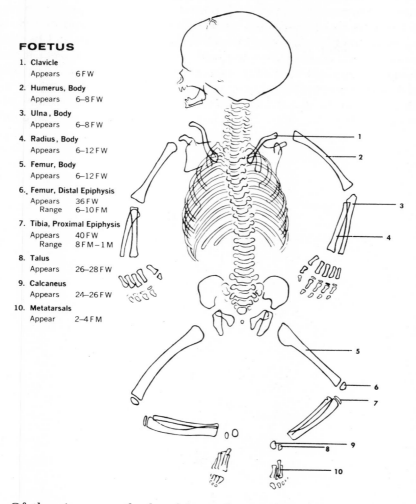

FOETUS

1. Clavicle
   Appears   6 F W

2. Humerus, Body
   Appears   6–8 F W

3. Ulna , Body
   Appears   6–8 F W

4. Radius , Body
   Appears   6–12 F W

5. Femur, Body
   Appears   6–12 F W

6. Femur, Distal Epiphysis
   Appears   36 F W
   Range   6–10 F M

7. Tibia, Proximal Epiphysis
   Appears   40 F W
   Range   8 F M – 1 M

8. Talus
   Appears   26–28 F W

9. Calcaneus
   Appears   24–26 F W

10. Metatarsals
    Appear   2–4 F M

Of the six areas, the hand is used most frequently in studies of child development. Many investigators take x-ray pictures of the hand only, for several good reasons. First, the hand is fairly representative of the total maturational status of the skeleton. Second, it contains a greater number of bony centers than any other joint area in the arms and legs; consequently, many maturational events can be observed and studied throughout the childhood period. Finally, the hand is an easy area to x-ray. The diagram, traced from the *Radiographic Atlas of Skeletal Development of the Hand and Wrist* pre-

pared by Drs. Greulich and Pyle, shows the bones of the hand and
wrist and the sequence in which these bones usually appear. Thirty
bony centers make their appearance after birth. Fourteen of these are
located in the fingers, six in the palm, eight in the wrist, and two at
the ends of the long bones in the forearm. The numbers appear to be
arranged at random as if they were deliberately scrambled. Characteristically, however, the bones appear in this particular sequence,
rather than in an ordered arrangement according to the regions of
the hand.   The diagram is a tracing of the bones of the hand of

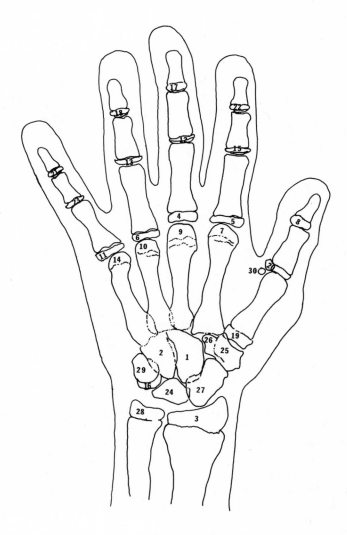

a boy approximately 14 years of age. At this time, the bones are well along in their developmental careers.

The age at which each bone appears is different for boys and for girls. The first bone, the capitate, appears around 2 months of age in girls and about a half month later in boys, while the thirtieth bone, the adductor sesamoid, does not appear until about 10½ years of age in girls and 12½ years of age in boys. The bones do not appear at equal intervals of time; further, each bone appears earlier in girls than in boys.

Why are the long bones in the fingers, the palm, and at the base of the wrist unnumbered? As you have probably guessed, these bones appear prenatally.

The Life Career of a Bony Center

Each bony center in the hand and wrist has its own life career, a career consisting of many maturational events, occurring in a fairly fixed sequence. In order to demonstrate the maturational process we need not laboriously trace the development of all 30 of the bony centers in the hand and wrist. We can effectively illustrate the process by studying the life career of a single bony center. The center selected for study is the third to appear. It is called the *distal epiphysis of the radius*. The name describes the exact location of the bone. The

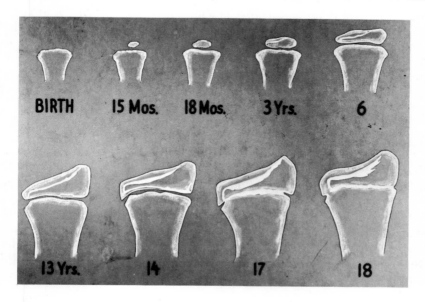

radius is the long bone in the forearm on the thumb side. It has two ends: a proximal end, which is closest to the midline of the body, and a distal end, which is farthest away. Since bone number 3 will eventually fuse with the shaft of the radius, it is called an *epiphysis*. A bony center that forms at the end of a long bone and ultimately fuses with it is called an epiphysis of that long bone.

In the enlarged diagram the life career of this bony center for a typical male has been summarized. Note the appearance of the radius at birth. The epiphysis is not in the picture simply because it has not yet begun to ossify. At approximately 13 months the epiphysis makes its appearance as a small calcified nubbin of bone. By following this bony center through successive ages, changes in form and shape can be observed; new margins and contours make their appearance sequentially. At approximately 17 years of age in the male, the epiphysis begins to fuse to the shaft. By 18, fusion is complete and the epiphysis has lost its identity as a separate bony center. Henceforth, its life career is merged with that of the radius as a whole. But maturation of this bone is not at an end. At about 35 to 40 years of age, arthritic lipping begins to occur around the edge of the radius where it joins the wrist bones. In old age, there will be a progressive loss of blood supply and changes in the texture and composition of the bone tissue; the bone will lose its resiliency and become brittle.

Thus far, we have discussed the *maturation* of the skeleton. It has been stated that maturation is concerned with the initial appearance of bony centers and their qualitative change through time. The number of independent bones present in the skeleton and the shape of each of the bones provide a yardstick for judging the maturational level of the individual. From 6 weeks after conception through very old age, the maturational stages of each and every bone in the skeleton are predictable and determinable.

Growth

Maturation is only one component in the story of the development of the skeleton. The other component is *growth*. When the skeleton of a newborn infant is contrasted with that of a young adult, the tremendous increase in size is readily apparent. At this point, we must reintroduce the concept of differential growth, discussed earlier. This refers to the fact that two or more parts of the body may be growing at different rates. If all six units of the skeleton grew at the same rate throughout the childhood period, the body would retain the proportions of the newborn infant right into adulthood. Actually, each unit has its own rate of growth so that the different parts of

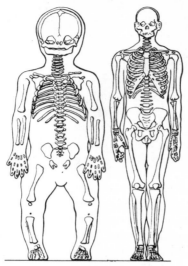

the body contribute *unequally* to stature at various points in time. One might imagine that because of differential growth of the individual bones a disharmonious skeleton would result, but the opposite is true. The differential growth of the skeleton results in a characteristic set of proportions for a particular level of development. The bodily proportions that are appropriate at one stage of development would be grotesque at another.

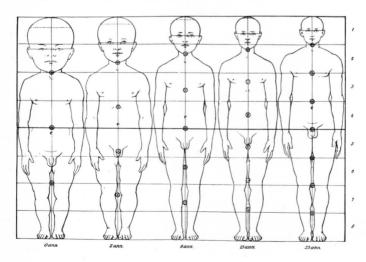

Growth in Length of a Long Bone

The growth of the skeleton as a whole is a function of the enlargement of each and every one of the hundreds of individual bony centers. In order to understand growth in stature, therefore, we must investigate the mechanisms of growth in the individual bone. The task then is *to grow a bone.*

We shall grow a long bone comparable to those found in the arm or leg. Approximately 5 to 6 weeks after conception, the long bone comes into existence, but it is very unlike the final bones of the adult skeleton. In fact, one might even say that the bone at its inception is not even a bone at all, if by bone we mean the hard, calcified material that we usually associate with the name "bone."

When the long bone first differentiates out of the tissue from which it is formed, it consists completely of cartilage. Cartilage, popularly called gristle, is a white, glistening, rubbery substance that we find, for example, in the movable part of our ears and in the tips of our noses. The cartilage does not become transformed into the final bone tissue; however, it is the framework on which the ultimate bone is built. In this sense, it is analogous to an inner wooden scaffolding around which a brick building is constructed. The cartilage forms a temporary, faithful model of the final bone.

Soon after the cartilage has taken the shape of a long bone, at about 2 months after conception, a ring of true bone tissue is formed around the middle of it, as shown in the picture. During the re-

mainder of prenatal life, the ring becomes wider and wider and extends toward both ends of the cartilage bone, until it covers the

entire length of the bone except the end portions. The ring has now become a sleeve surrounding the cartilage bone.

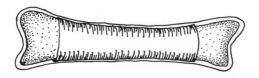

As the sleeve extends toward both ends of the long bone, the cartilage encircled by the widening sleeve is torn down and removed as waste material by the bloodstream. When the cartilage is removed, it is replaced by spongy bone. Both the sleeve and the spongy bone that replaces the cartilage within the sleeve are true bone tissue called *osseous tissue*. The osseous tissue is the hard substance we see when we examine a bone.

At the time of birth, the long bone consists of a tubular sleeve filled with spongy bone; it possesses solid cartilage extensions at the ends of the sleeve. These cartilage extensions are essential to the further growth of the bone; in fact, they are the mechanism that enables the bone to grow in length. The long bone can grow in length *only* at the cartilage ends. *The growth of the cartilage, not the osseous tissue, actually accomplishes the lengthening of the bone.* Therefore, the replacement of the cartilage by the osseous tissue is a secondary event. Thus, a long bone grows not as a result of growth of bone but as a result of growth of cartilage.

If you examine an x-ray of a long bone of a newborn infant, you will see the osseous tissue but not the cartilage extensions at the ends. The cartilage extensions do not show up on the x-ray because cartilaginous tissue is not dense enough to cast a shadow. X-rays pass right through like sunlight through window glass.

As the cartilage extensions continue to grow at the ends of the long bone, the osseous sleeve will be added to at its ends. Rapid formation of cartilage cells at the ends of the bone brings about the exuberant growth of the childhood years.

The next step in the growth of the long bone is an event with which we are already familiar, the appearance of a bony epiphysis in the cartilage extension at each end of the bone. First evident as a

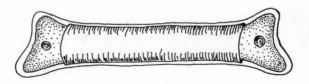

small nubbin in the center of the cartilage extension, the epiphysis grows larger and wider, replacing the cartilage until it is as wide as

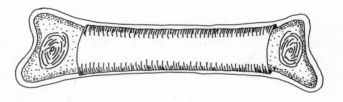

the end of the sleeve, the tubular bony shaft. When this point is reached, the cartilage that originally formed the entire end of the bone becomes reduced to a thin disc sandwiched between the bony epiphysis and the shaft. During the childhood years the disc remains

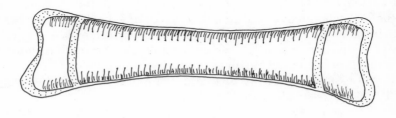

at a fairly constant thickness, actively growing cartilage cells that are replaced by the spongy bone on the inside and surrounded by the lengthening sleeve on the outside. This disc now carries the grandiose label *cartilaginous epiphyseal disc*.

Who would suspect this inner drama of creation and death unfolding in the apparently inert bones of the growing child? If we wish to wax romantic, we might even say that the good work of the cartilage is an exercise in frustration. Just as quickly as it grows, it is torn down and replaced by osseous tissue.

Eventually, the cartilage disc begins to slow down in its rate of growth. But the inexorable replacement by osseous tissue continues unabated, and the disc becomes thinner and thinner. Ultimately,

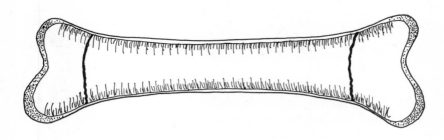

the cartilage gives up the ghost and stops growing entirely. It is completely replaced by the compact bone of the sleeve on the outside and the spongy bone on the inside. When this happens, the bony epiphysis and the bony shaft meet and the two fuse together. This process, of course, occurs at both ends of the shaft, though not necessarily at the same time. The physiological mechanism of growth, the cartilage discs, has been eliminated; the bone will not undergo further growth in length. The event that brings the growth of the long bone to an end is a maturational event. Thus, maturation actually controls the duration of growth.

Growth in Thickness

So far we have discussed the process by which a long bone becomes longer. Obviously, a long bone also grows in thickness. What is not so obvious is the way in which this is accomplished. The mechanism controlling the growth of the bone in thickness is much simpler than the cartilage replacement mechanism responsible for growth of the bone in length.

A long bone grows in thickness in the same way that a tree trunk increases in circumference. New rings of bone are added in circular fashion outside the old layers. If this process continued unchecked, the compact shell of the bone would increase enormously in thickness. But the bone has a compensating mechanism that controls the thickness of the shell. As new rings are added on the outside (2, 3, 4), old layers are removed from the inner circumference of the sleeve (0, 1), thus permitting the shell to become thicker

but not excessively so. This process continues until each of the long bones reaches full size in late childhood or early adulthood.

During the early childhood period the tubular portion of the bone is filled with spongy bone tissue. In later childhood, as the bone approaches final size, the spongy bone within the tube is gradually removed, and by adulthood the bone is hollow for most of its length. The spongy bone persists only at the ends.

Modeling Resorption

We have discussed how a long bone grows in length and in breadth, but another important feature remains to be described. At birth, the bone tends to flare out at its ends and to be thinner in the middle. As the bone grows in length, the end of the bone becomes wider and the shaft becomes rather heavy and clublike. Yet when we look at an adult bone, it has a rather graceful appearance. Nature has a mechanism for maintaining the slimness of the bone. This mechanism is called *modeling resorption*. As the ends of the bone become excessively thick, the excess bone tissue is removed by a process of resorption or breakdown. This process permits the bone to maintain its slim shape while it is growing longer.

When the resorption mechanism fails, a distorted and misshapen bone results. The deficiency of the remodeling apparatus is a characteristic phenomenon in the grey-lethal and microphthalmic mutants of the mouse. A comparison of the cheekbone in a normal mouse with a specimen of each of these strains provides a dramatic illustration of the importance of the remodeling mechanism for the retention of the normal shape of the bone during development.

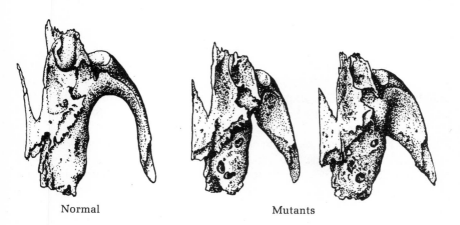

Normal   Mutants

Staining the Growing Bone

A bone can *only* become larger by having new layers of bone added to old layers, either at the ends of the bone or around the circumference. In other words, a bone grows like a brick wall, new building blocks are added onto old. This mode of growth is called *appositional growth* because new layers are added or apposed to old.

We know that bones grow in this fashion from the experimental study of living animals. Research in this area was significantly advanced by the accidental discovery of a novel method for studying the growing bone. This discovery had its origin in a very mundane activity. Dr. John Belchier, an English surgeon, was dining at a friend's house in the year 1736. To his amazement he noticed that the bones of the pork he was eating were stained red. His friend, by vocation, was a dyer who used the red juices of the *madder plant* to fix the dyes in his cloth. Pursuing the matter further, Dr. Belchier discovered that his friend used bran to absorb the excess dye when he cleaned the vats. Being a thrifty man, he was reluctant to discard the madder-stained bran, so he fed it to his pigs. With this information, the mystery was solved. The juice of the madder plant eaten by the pigs had stained their *growing bones* red.

From this rather curious beginning, over 200 years ago, came the technique of *vital staining,* the staining of the growing bones in the living animal. This technique is used to the present day with remarkably little change in basic procedure. Since the initial experiments, scores of unsuspecting pigs, sheep, rats, rabbits, dogs, cats, and monkeys have spent a considerable portion of their growing years with colored bones. Only that part of the bone growing at the time the dye is absorbed is stained red. The bone formed prior to the period of staining, as well as that laid down subsequently, remains white. The application of the vital staining method has increased our knowledge of the mechanisms of bone growth.

In this section we have discussed the two mechanisms controlling the growth of the long bone in the arm or leg. One mechanism is responsible for the growth of the bone in length, the other increases the bone's thickness or circumference. At its inception, the long bone consists entirely of cartilage. The growth of this cartilage is *solely* responsible for growth of the bone in length; osseous tissue simply replaces the cartilage. The maturational event of fusion of the shaft with the epiphysis eliminates the cartilage growth mechanism, thereby terminating growth of the bone in length. When all of the epiphyses in all of the bones contributing to stature have fused with their shafts, growth in stature is complete.

Statural growth is a deceptively complex phenomenon that requires an understanding of skeletal development. But the implications of skeletal development extend far beyond stature. We shall see that the study of the skeleton assists in predicting a child's ultimate adult stature. It provides a record of a child's health history and can be used to predict when a girl will begin menstruating. From the skeleton alone, we can determine a person's race and sex and his age at death. In other words, the skeleton serves as a window through which we may see many aspects of the development of the individual.

Chapter 5

MORE ABOUT BONES

Each of us, man or woman, boy or girl, has two ages: a chronological age, which tells how long we have lived, and a biological age, which expresses our actual achievement on the pathway of development. Biological age is a recognition of individual differences in the rate of development. When biological age is well in advance of chronological age, the individual is developing rapidly; when the reverse is true, the individual is exhibiting a slow rate of development.

Determining Skeletal Age

Biological age represents an aggregate or composite of many discrete developmental factors. Ideally, all the systems and organs of the body would be assessed to determine biological age. Due to the enormity of the task, however, in actual practice biological age usually incorporates only certain key measures of development, such as height, weight, skeletal maturation, dental eruption, and the degree of development of secondary sex characteristics. Since each of these measures is different in its basic nature, in order to make direct comparisons a common denominator is required. For each of the measures of development the raw scores must be converted into a single system of biological age. For example, given a 7-year-old boy whose height is 50 inches and whose weight is 57 pounds, we must convert the height, which is a linear measure, and the weight, which is a measure of mass, into a single system. Raw scores are converted into age equivalents by determining the average age at which boys achieve a height of 50 inches and the average age at which they achieve a weight of 57 pounds. Referring again to the example, the boy's height, converted to height age, is 8 years. His weight converted to weight age is also 8 years. Thus, in height and weight this boy is developmentally a year in advance of his chronological age.

By substituting an age equivalent, it is possible to convert the raw

77

measurements for each bodily trait into a biological or developmental age. The different bodily traits may then be compared directly for the purpose of determining whether the child is advanced in some areas and lagging in others. When the developmental ages for all the bodily traits are assessed and averaged, a single biological age representing a composite picture of the child's level of development is obtained. Biological age is sometimes called "organismic age," implying that the total organism can be summed up in one composite score.

The specific developmental factor that is the primary concern in this chapter is *skeletal age*, the age assigned to an individual skeleton at a particular point in its development. It is a recognition of the fact that individuals of comparable chronological age may differ widely in their level of skeletal maturity. By comparing skeletal maturity with chronological age, we can determine whether the child is maturing rapidly or slowly.

Before assessing a child's skeletal development, we must first establish a set of standards typical of children at each age. Assuming that such standards were not available, we would begin by taking x-rays of a large sample of boys (or girls) at birth. These x-rays would be placed on a table long enough to permit us to spread them out in a single row in order to find the x-ray most characteristic or typical of the entire collection of x-rays. This task is accomplished in several stages. Since each of the bones in the skeleton undergoes a series of maturational changes that are identifiable, the x-rays would be arranged in a sequence of maturity, from least mature at one end of the row to most mature at the other. To locate the typical x-ray for the age group in question, the extremes at both ends must be discarded. The procedure of eliminating the extreme x-rays is repeated again and again; each time the remaining x-rays represent a reduced range of difference. Finally, a small group is left. This is called the modal group. Through a process of careful reappraisal, the most representative x-ray is selected. This x-ray now becomes the standard bearer of the skeletons of boys at birth; it becomes the prototype for determining skeletal age in newborn boys. Any boy whose x-ray matches the level of maturity represented by the standard is assigned the skeletal age "newborn," irrespective of the boy's actual chronological age. In order to obtain standards for subsequent chronological ages, the same procedure is repeated when the boys are 3 months old; then every 3 months through the age of 2 years, and every 6 months thereafter. Ultimately, a series of standards would be established from birth through young adulthood.

Fortunately it is not necessary to construct our own standards. Several published atlases are available which photographically depict the most typical levels of skeletal maturity for each chronological age.

Since the latter part of the nineteenth century, with the invention of the x-ray machine, anatomists have been mapping the life careers of all the bony centers in the body. They have ben studying the time at which these centers appear, the sequence of their appearance, and their rate of maturation in different populations and races. Atlases have been constructed for different areas of the skeleton, showing the progression from initial appearance through the last stage of maturation for each bony center.

Since the atlases depict the typical or model rate of skeletal maturation for the population studied, we are confronted by a problem. Only the rare child matches exactly the standard for his age. Most children fall in between standards; therefore, we often assign a skeletal age half-way between two standards. Or, if we are proficient in gauging the subtle differences between a particular child's x-ray and the standards in the atlas, we may say that he is 1 or 2 months away from a given standard. These procedures may be illustrated by applying them to a specific child. Let us determine a skeletal age for Jim Jones at 7 years of age. An x-ray of his hand and wrist is reproduced below. In assessing his skeletal age, we shall use the *Radiographic Atlas of Skeletal Development of the Hand and Wrist*, compiled by Drs. William W. Greulich and S. Idell Pyle. Next to his x-ray are the standards from the Atlas representing skeletal ages 6 and 7. A comparison of Jim's x-ray with the standards shows that it falls between the 6- and the 7-year-old standards. Further examination indicates that Jim's x-ray represents a level of development closer to the 6-year-old standard than the 7-year-old. We have, therefore, assigned him a skeletal age (SA) of 6 years, 4 months. In other words, he is 4 months beyond the 6-year-old standard and 8 months below his chronological age.

Is the discrepancy between Jim's chronological age and his skeletal age significant? Is he retarded in his development? The answer to both of these questions is "no" because two out of every three boys Jim's age deviate by as much as 9 months on either side of the 7-year-old standard. Since Jim is within this range, we conclude that he is a normal child following a slow schedule of development. Even if Jim had deviated more than 9 months from the standard for his age, we would not be immediately alarmed. We would want to study him for a period of time. Only if we found a changing rate of development

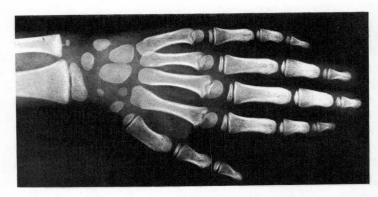

MALE STANDARD - 15

S A - 6 YEARS

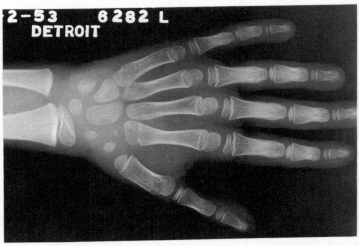

JIMMIE JONES

C A - 7 YEARS

S A - 6 YEARS, 4 MONTHS

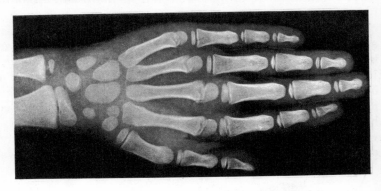

MALE STANDARD - 16

S A - 7 YEARS

that carried him further and further away from the standard for his
age would we conclude that Jim was not making satisfactory progress.
Our assessment of the level of skeletal maturity of this boy was

Name___ Jones, Jimmy_____ Birthdate___ 1-26-46_____ X-ray No. ___ 6282_____

	Age Equivalents
Date of x-ray	2-2-53
Chronological Age	7-0-6
Distal End of Radius	84
Distal End of Ulna	(63)
Capitate	84
Hamate	84
Triquetral	84
Lunate	72
Navicular	84
Greater Multangular	72
Lesser Multangular	67
Metacarpal I	84
Metacarpal II	78
Metacarpal III	78
Metacarpal IV	78
Metacarpal V	78
Proximal Phalanx I	72
Proximal Phalanx II	72
Proximal Phalanx III	72
Proximal Phalanx IV	72
Proximal Phalanx V	72
Middle Phalanx II	72
Middle Phalanx III	72
Middle Phalanx IV	72
Middle Phalanx V	72
Distal Phalanx I	84
Distal Phalanx II	76
Distal Phalanx III	76
Distal Phalanx IV	76
Distal Phalanx V	72
Pisiform	*
Adductor Sesamoid (Thumb)	*
Flecor Sesamoid (Thumb)	*
Skeletal Age (Mean)	76
Most Advanced bone	84
Least Advanced bone	63

based on a general inspection of the bones in his hand and wrist as a
total group. But it is the rare child whose bones are all developing at
the same rate and can be assigned the same skeletal age. In fact, most
children differ appreciably in the rate of development of their individual bones. In actual practice, therefore, the inspection technique
of matching the whole hand is supplemented by a more systematic
and accurate method of determining skeletal age. This method consists of assigning a separate skeletal age to each of the bony centers
and then averaging all of these to get the representative skeletal age
for the entire hand. We have followed this procedure for Jim Jones,
and have assigned individual ages to each of the bones of his hand
and wrist. The assessment shows a spread in age from 5 years, 3
months to 7 years, 0 months. The average age of his individual bones
is 6 years, 4 months, the same as the skeletal age assigned by the
inspectional method.

It is often useful to study the spread of skeletal ages assigned to
the individual bones. Marked variations may simply reflect a familial
pattern, but they may also suggest nutritional deprivation. Children
reared on unsatisfactory diets, particularly those deficient in calcium,
show considerable differences among the ages of their bones. Studies
indicate that when the diet is enriched and supplemented with ade-
quate quantities of milk, children improve dramatically. The bones
most retarded in rate of development show a pronounced tendency
to catch up.

When we assess a child's skeletal maturity, as we have done with
Jim Jones, we determine his *status* at one point in time. The assessment indicates whether he is more or less mature than his age peers
at that particular time. But it is a static view. Only very conservative
judgments or predictions should be made from a single assessment
of a child's status. What is needed is a picture of his development
over a period of time, a picture of his development *progress*. From
this, we can tell how rapidly or how slowly the child is moving along
his pathway to adult maturity. To illustrate this point, we shall consider the x-rays of Jim Jones once again, looking at his skeletal age
at annual intervals from birth through his present age of 10 years.
The ten skeletal ages have been plotted on a graph, permitting us to
assess his progress at a glance. Chronological age is plotted on the
horizontal axis and skeletal age on the vertical axis. The diagonal
straight line running from the lower left to the upper right corner is
the path a child follows if his skeletal age is always the same as his
chronological age. When a child's skeletal age is in advance of his

chronological age, his curve rises above the diagonal line; when it is behind chronological age the curve falls below the diagonal line. In Jim Jones's case the curve shows a fairly consistent and regular pattern slightly below the standard of reference, the diagonal line. All

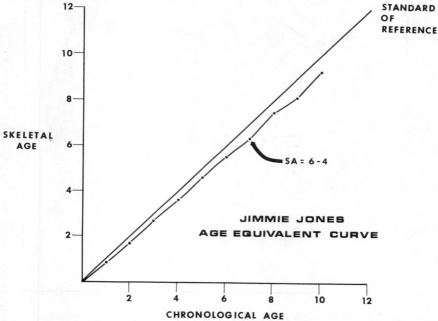

along the path he has been fairly consistent in his skeletal maturation. The single "reading" we obtained when he was 7 is not an aberrant observation but fits the pattern of his progress during his first 10 years of life. He is developing in a consistent way within the normal pattern but slightly slower than average.

Thus far in our discussion we purposely did not lump the girls with the boys for a very significant reason—girls develop more rapidly than boys. Even at birth most girls are more mature than boys, and they stay that way throughout childhood. Consequently, it is necessary to construct a separate set of standards for girls. When the x-rays of boys and girls in the atlases are compared, girls consistently are found to be one or two standards ahead of boys of the same chronological age. In the illustration, the skeletal age standards (SA), taken from the Greulich and Pyle Atlas, have been enlarged to the same size in order to highlight the fact that the level of maturation is comparable although there is a 2-year differential in age.

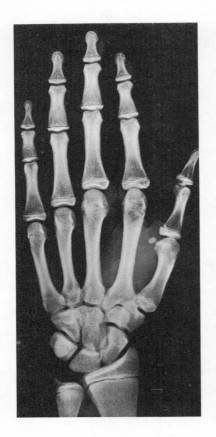 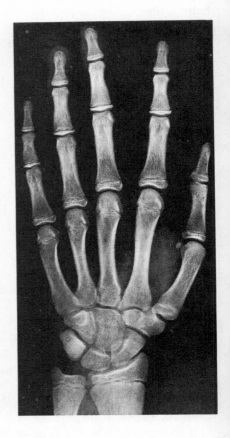

MALE   FEMALE

STANDARD 27   STANDARD 22

SA : 15y, 6m   SA : 13y, 6m

Predicting Adult Stature

The concept of skeletal maturation can be used as a working tool to reveal certain significant facts concerning development. The first of these is the prediction of adult stature during the childhood years. To make this prediction three items are required: the child's chronological age, his skeletal age, and his present stature. The method may be illustrated with a concrete example. Ricky K., at age 13, was appreciably shorter than his peers. His stature was 55 inches. From a bone-by-bone analysis of his hand and wrist x-ray, it was determined that he had a skeletal age of 11 years, 9 months. In terms of *growth* he was a small boy, 5½ inches shorter than the average boy his age; in terms of *maturation* he was more than a year behind his chronological age. Does this mean that Ricky will be unusually short as an adult? By estimating his adult stature and comparing his predicted height with the average of young adults, this question can be answered. Our task is facilitated by the Bayley and Pinneau tables for predicting adult stature. These tables were designed for use with the Greulich and Pyle Atlas. Although the table we have reproduced resembles an actuarial chart from an insurance company, it is actu-

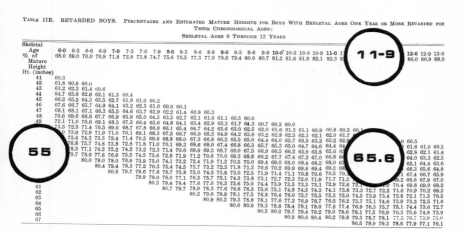

TABLE IIE. RETARDED BOYS. PERCENTAGES AND ESTIMATED MATURE HEIGHTS FOR BOYS WITH SKELETAL AGES ONE YEAR OR MORE RETARDED FOR THEIR CHRONOLOGICAL AGES; SKELETAL AGES 6 THROUGH 13 YEARS

Skeletal Age	6-0	6-3	6-6	6-9	7-0	7-3	7-6	7-9	8-0	8-3	8-6	8-9	9-0	9-3	9-6	9-9	10-0	10-3	10-6	10-9	11-0	…	11-9	…	12-6	12-9	13-0
% of Mature Height	68.0	69.0	70.0	70.9	71.8	72.8	73.8	74.7	75.6	76.5	77.3	77.9	78.6	79.4	80.0	80.7	81.2	81.6	81.9	82.1	82.3		*(11-9)*		86.0	86.9	88.0
Ht. (inches)																											
41	60.3																										
42	61.8	60.9	60.0																								
43	63.2	62.3	61.4	60.6																							
44	64.7	63.8	62.9	62.1	61.3	60.4																					
45	66.2	65.2	64.3	63.5	62.7	61.8	61.0	60.2																			
46	67.6	66.7	65.7	64.9	64.1	63.2	62.3	61.6	60.8	60.1																	
47	69.1	68.1	67.1	66.3	65.5	64.6	63.7	62.9	62.2	61.4	60.8	60.3															
48	70.6	69.6	68.6	67.7	66.9	65.9	65.0	64.3	63.5	62.7	62.1	61.6	61.1	60.5	60.0												
49	72.1	71.0	70.0	69.1	68.3	67.3	66.4	65.6	64.8	64.1	63.4	62.9	62.3	61.7	61.3	60.7	60.3	60.0									
	73.5	72.5	71.4	70.5	69.6	68.7	67.8	66.9	66.1	65.4	64.7	64.2	63.6	63.0	62.5	62.0	61.6	61.3	61.1	60.9	60.8	60.5	60.1				
	73.9	72.9	71.9	71.0	70.1	69.1	68.3	67.5	66.7	66.0	65.5	64.9	64.2	63.8	63.2	62.5	62.3	62.1	61.7								
	75.4	74.3	73.3	72.4	71.4	70.5	69.6	68.8	68.0	67.3	66.8	66.2	65.5	65.0	64.4	64.0	63.7	63.5	63.3	63.2	62.9				60.5		
	76.8	75.7	74.8	73.8	72.8	71.8	71.0	70.1	69.3	68.6	68.0	67.4	66.8	66.3	65.7	65.3	65.0	64.7	64.6	64.4	64				61.6	61.0	60.2
	78.3	77.1	76.2	75.2	74.2	73.2	72.3	71.4	70.6	69.9	69.3	68.7	68.0	67.5	66.9	66.5	66.2	65.9	65.8	65.6	65				62.8	62.1	61.4
	79.7	78.6	77.6	76.6	75.5	74.5	73.6	72.8	71.9	71.2	70.6	70.0	69.3	68.8	68.2	67.7	67.4	67.2	67.0	66.8	66				64.0	63.3	62.5
		80.0	79.0	78.0	76.9	75.9	75.0	74.1	73.2	72.4	71.9	71.2	70.5	70.0	69.4	69.0	68.6	68.2	68.0	67					65.1	64.4	63.6
			80.4	79.4	78.3	77.2	76.3	75.4	74.5	73.7	73.2	72.5	71.8	71.3	70.6	70.2	69.9	69.6	69.4	69.3	68				66.3	65.6	64.8
				80.8	79.7	78.6	77.6	76.7	75.8	75.0	74.5	73.8	73.0	72.5	71.9	71.4	71.1	70.8	70.6	70.5	70				67.4	66.7	65.9
					79.9	79.0	78.0	77.1	76.3	75.7	75.1	74.3	73.8	73.1	72.7	72.3	72.0	71.9	71.7	71.3				9.2	68.6	67.9	67.0
					80.3	79.4	78.4	77.6	77.0	76.3	75.6	75.0	74.4	73.9	73.5	73.3	73.1	72.9	72.6	72.7				70.4	69.8	69.0	68.2
61						80.7	79.7	78.9	78.3	77.6	76.8	76.3	75.6	75.1	74.8	74.5	74.3	74.1	73.8	73.3	72.7	72.2	71.6	70.9	70.2	69.3
62							80.9	80.2	79.3	78.8	78.1	77.6	76.9	76.7	76.5	76.2	75.7	75.1	74.6	73.9	73.3	72.5	71.6			
63								80.2	79.6	78.9	78.1	77.5	76.8	76.4	76.0	75.7	75.5	75.3	75.0	74.5	73.9	73.4	72.8	72.1	71.3	70.5
64									80.5	80.0	79.3	78.8	78.4	78.1	78.0	77.8	77.4	76.9	76.3	75.7	75.1	74.4	73.6	72.7		
65										80.6	80.0	79.3	78.8	78.4	78.1	78.0	77.8	77.4	76.9	76.3	75.7	75.1	74.4	73.6	72.7	
66											80.5	80.0	79.7	79.4	79.2	79.0	78.6	78.1	77.6	76.9	76.3	75.6	74.8	73.9		
67												80.9	80.6	80.4	80.2	79.8	79.3	78.7	78.1	77.5	76.7	75.9	75.0			
													80.5	79.9	79.3	78.6	77.9	77.1	76.1							

(Handwritten circled annotations on the table: "11-9", "55", and "65.6".)

ally the table for boys whose skeletal age is more than a year behind their chronological age. This is the relevant table for predicting Ricky's adult stature. First, it is necessary to locate his present height (55 inches) as indicated in the column on the left. Then, reading across the top of the chart, we find his skeletal age (11 years, 9 months). At the point of intersection of his present height with his skeletal age, we obtain his predicted adult stature—65.6 inches. At 15, Ricky was x-rayed and measured again and a new prediction was made. His stature at that time was 61 inches, 4 inches shorter than the average of the population. His skeletal age was 14.0 years, a year behind his chronological age. Turning again to the Bayley-Pinneau tables, a predicted adult stature of 65.8 inches was obtained. This prediction is remarkably similar to the earlier one. His estimated adult stature at age 15 would place him only 2½ inches below the average. By graphing the material, Ricky's changing status and his progress can be easily visualized. His actual stature at each chronological age has been plotted in relation to the average curve of growth for boys. Since his curve is steeper than the average, we can reassure him that he is closing the gap. As the graph indicates, he did not reach his terminal stature until 20 years of age.

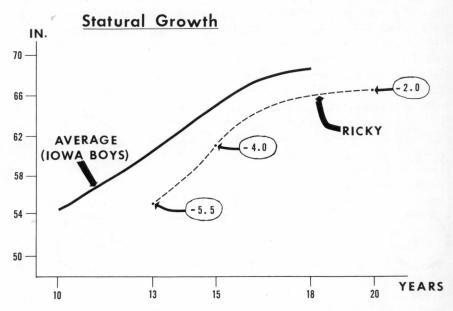

At young adulthood, Ricky's stature was 66.5 inches, approximately ¾ of an inch more than our prediction. Although Ricky did not reach the average height of young adults, he made a distinct improvement in status. He was a slowly maturing boy who reached his adolescent growth spurt somewhat later than the average. Knowledge of his pattern of development enabled us to predict his adult stature with a degree of accuracy that could not have been accomplished from statural measurements alone.

Ricky's pattern of development demonstrates an important principle: *The less mature the skeleton, the more growth potential remains.* When the skeleton is immature, the long bones are still growing; they will continue to grow until fusion takes place. Thus, the rate of skeletal maturation influences eventual adult stature. Ricky's skeletal immaturity was his greatest asset because it permitted continued growth.

Predicting the First Menstruation

For girls, in addition to predicting adult stature from skeletal age, we can also predict *the time of menarche,* the first menstruation. The *average* girl begins menstruating at about the time she reaches a skeletal age of 13 years, 5 months, *irrespective of her chronological age.* Two out of every three girls begin menstruating within plus or minus 5 months of this time. In terms of chronological age, however, these same girls would show a spread of 20 months at menarche. Therefore, by knowing skeletal age, we can predict when a girl will begin menstruating with twice the accuracy that is possible from knowing her chronological age alone.

The *early-maturing* girl will have a skeletal age in advance of her chronological age and will begin menstruating early. The *late-maturing* girl, on the other hand, will have a skeletal age younger than her chronological age, and she will begin menstruating late. Some girls begin menstruating as early as 10 years of age, while others do not begin menstruating until 15 years of age. There are even rare instances on record in which the first menstruation occurs in infancy, as early as 7 months of age. It is interesting to note that these unfortunate girls also have tremendously advanced skeletons.

Prior knowledge of the ultimate adult stature and the time of menarche provides us with valuable aids in counseling early- and late-maturing girls. The late-maturing girl who suffers embarrassment because of delayed sexual development can be reassured that she is progressing normally and will reach sexual maturity in accordance with her own timetable. In terms of her present short stature, we can offer her the additional support that late maturing girls generally grow for a longer time and are frequently taller adults than early maturing girls. Assurance can also be given to the early-maturing girl that despite her rapid pattern of development, she will not grow to be a giant. Her growth will cease rather abruptly when she reaches sexual maturity, at a much earlier age than her peers.

The dramatic effect of the timing of menarche on growth in stature may be summarized in a single illustration. The graph depicts the curves of growth for twenty girls, ten early maturers who first menstruated between 10 years, 7 months, and 11 years, 3 months, and ten late maturers who reached menarche between 14 years, 5 months, and 15 years, 2 months.

The illustration highlights the fact that the early-maturing girls are on a faster timetable than the late-maturing girls. *On the average,*

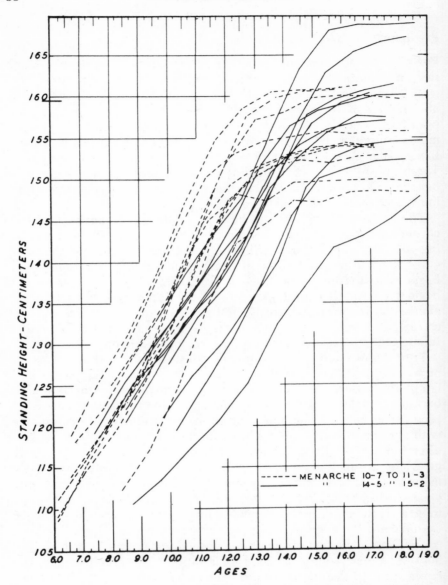

they are also bigger in childhood. They go through the whole course of growth earlier and cease growing abruptly. Conversely, the slow-maturing girls are smaller in early childhood, continue to grow for a longer period of time, and ultimately achieve greater stature than early maturers. Of course, individual variation is evidenced in both groups. Some early maturers continue to be taller into adulthood than some of the late maturers, while some late maturers continue to be shorter. The curves of the twenty girls in the above illustration appear to be markedly different in form. However, when the curves for the early maturers are superimposed on the average size for that age group, and when this is also done for the late maturers, a remarkable similarity in pattern of growth is revealed.

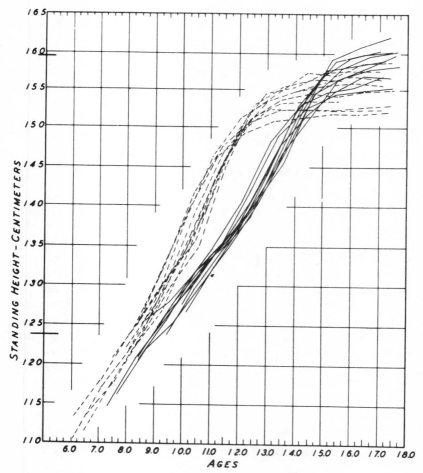

In our technologically oriented society, we are loathe to accept, fatalistically, that we will not achieve the cultural ideal in height. To date, medical science has not discovered a method to make boys taller or girls shorter. Attempts to use different hormones have had the undesirable effect of stimulating the *maturation* of the skeleton, as well as its *growth*. Most studies indicate that boys treated with male sex hormones for the purpose of stimulating growth reached their terminal adult heights sooner but were no taller than was predicted for them prior to treatment.

In our earlier discussion we stated that the more mature the skeleton, the less growth potential remains. Early maturation brings a terminus to growth in stature. Endocrinologists are attempting to discover a hormonal substance, either natural or synthetic, which when administered to a child will stimulate his *growth* but not accelerate his *maturation*. But still the most certain way for a child to insure himself of tall stature is to select tall parents. When statures of parents and children are compared, a consistently high correlation is obtained; tall parents beget tall children. Unfortunately, the child cannot select his genes from a supermarket shelf. However, an optimal environment of good nutrition and preventive medicine will enable the child to exploit his growth potential to the maximum.

The Skeleton Records Past History

Thus far, we have used the skeleton to foretell future events in the child's development. The skeleton, however, is also a recorder of things past. From the skeleton, the health history of the individual can be reconstructed. During childhood, for example, severe illnesses are often recorded in the long bones of the skeleton. This record can be seen on an x-ray in the form of a series of white lines—called "lines of arrested growth"—at the growing ends of the long bones. It is believed that when a child suffers a severe upper respiratory infection or some other debilitating illness, the growth of the matrix of his bones is temporarily slowed down, but the deposition of calcium and phosphorus continues unabated. Consequently, the bone has a small zone of more heavily mineralized tissue, which appears on the x-ray as a white band. When the child recovers from his illness, the growth of the matrix resumes its normal rate and the subsequently formed tissue is once again normally mineralized. The end of the bone continues growing, but the line of arrested growth is left behind as a temporary record of the illness. As the bone grows and is reconstructed with new tissue, this line eventually disappears.

The accompanying picture illustrates lines of increased density in the end of the thighbone and both ends of the shinbone in the leg of a young girl. The lines that appeared at 18 months have been retained, while a new series of lines has been added by 24 months. The same

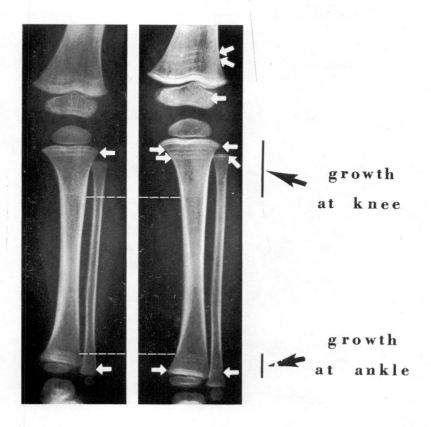

18 months   24 months

pattern of heavy and light lines may be seen in both bones of the leg. The amount of growth that occurred in the interval of 6 months is greater at the knee joint than at the ankle, thus illusrating differential growth even within a single bone.

Even physical disasters can be recorded in the skeleton. While assessing the effects of the atomic bombing of Nagasaki and Hiroshima, for example, scientists frequently encountered x-rays of children showing a single strong line of arrested growth in the long bones. They attributed this line to a temporary interruption of growth caused by the injuries sustained in the bombings.

The skeleton can also record parental solicitation. The well-intentioned efforts of parents sometimes introduce lines of denser mineralization into the bones of children. In the case of a child who received, each winter, large doses of cod liver oil, heavily fortified with phosphorus, an x-ray revealed the seasonal variation in the bone density. Since the bones were more heavily mineralized during the winter months, the x-ray exhibited rings reminiscent of those of a tree stump. In the picture below, the dramatic results are depicted.

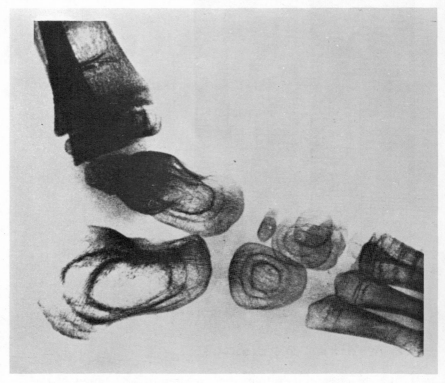

Literally, it would not be correct to call these lines of heavier mineral deposit "lines of arrested growth" because, in fact, growth has not been arrested. The lines simply reflect seasonal difference in the amount of mineral content supplied to the bones. The regularity of the pattern is in sharp contrast with that formed as a result of illness.

Functions of the Skeleton

We have examined the skeleton to uncover the past and to predict the future; now let us consider the present. How does the skeleton serve us today? What are its functions?

The most obvious function of the skeleton is that of support. The skeleton provides the *internal framework* of the body, which gives the organism its shape and proportions. In this sense, the skeleton is comparable to the structural steel framework of a skyscraper, built to withstand the stresses and strains imposed upon the building by its own weight. The skyscraper must also possess a measure of resiliency so that it can bend slightly under the buffeting of a high wind. The bones of the skeleton also evidence this attribute.

If you are conversant with the jargon of structural steel workers you know that they use I beams in the construction of buildings and bridges. The cross-section of the beam resembles the capital letter I. Beams of this shape are used because of their inherent strength and resistance to bending under compression or torsion. The bones of the skeleton are also designed to withstand these forces. The skeleton uses the principle of hollow tubular construction rather than the I-shaped beam. The hollow tube of the long bone is actually stronger than a solid bone containing the same amount of material would be.

If a long bone were sliced lengthwise, a system of internal buttresses would be revealed at both ends. The buttresses are designed to maintain the shape of the bone under stress and strain. The accompanying x-ray shows the radiating lines that form the internal structure of the head and neck of the thighbone. This internal buttress helps the bone maintain its shape and counteract the distorting effect

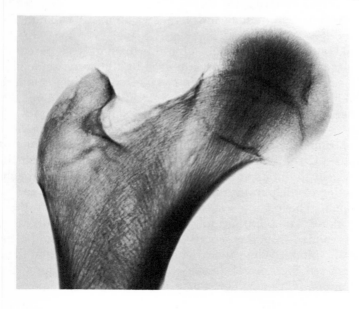

of the weight of the body. In this way, the bones of the skeleton erect for themselves an internal system of braces that help them maintain their shape in spite of pressures and strains, in the same way that a bridge builder places a series of diagonal steel braces between the main girders to help the bridge resist torsional effects that might destroy it.

When the weight-bearing function of the thighbone is eliminated, as in the case of amputation of a leg not replaced by an artificial limb, the internal buttressing structures of the bone atrophy and become thinner. The thighbone of Long John Silver would evidence degeneration because he had transferred the weight-bearing function to a crutch. Peg Leg Pete, on the other hand, avoided atrophy by use of a wooden leg. In this case, the thighbone continued to bear the weight of his body and thus the internal buttressing was maintained.

Although we can make an analogy between the skeleton and the framework of a building up to a point. Unlike a building, the skeleton facilitates mobility. It enables the person to move about easily, allowing arms, legs, head, neck, and back to move in space, thereby permitting the individual to perform myriad motions and actions.

A second function of the skeleton is its role as a banking institution. The currency of this bank is calcium, a vital element in the normal functioning of the body. An insufficient level of calcium in the bloodstream severely impairs the normal functioning of the nervous and muscular systems. For this reason, normal blood calcium levels must be maintained and a ready supply of calcium must be available.

The skeleton is the storehouse for approximately 98 percent of the calcium that can be withdrawn and used by the body, as needed. The body absorbs calcium from the food ingested and deposits a reserve in the skeleton. Then, like a banking institution, on demand the skeleton pays out the calcium to the bloodstream. When the level of the calcium in the blood falls to the danger point, a mechanism in the body instructs the outpaying teller's cage to take calcium from the storehouse of the bones and put it back into the blood so that the organism can function in a healthy way. This ingenious mechanism is a function of the parathyroid glands, which are situated in the front of the neck. These glands secrete a hormone, called parathormone, into the blood. The parathormone stimulates the liberation of calcium from the bones into the bloodstream, thereby raising the calcium level of the blood. When the calcium concentration in the blood has been restored to normal levels, the parathyroid glands shut off parathormone secretion. This sensitive mechanism is analagous to the thermostat in our homes, which operates as a feedback and sends an impulse to the furnace, turning it on when the temperature falls below the present level and turning it off when the temperature rises above the preset level.

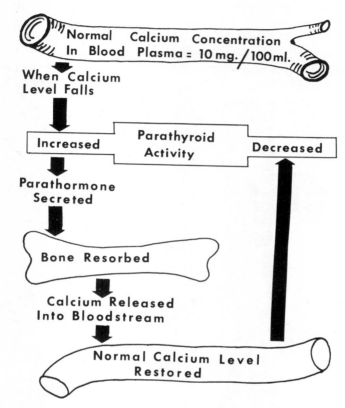

Normal Calcium Concentration
In Blood Plasma = 10 mg. / 100 ml.

When Calcium
Level Falls

Increased | Parathyroid Activity | Decreased

Parathormone
Secreted

Bone Resorbed

Calcium Released
Into Bloodstream

Normal Calcium Level
Restored

PARATHYROID FEEDBACK MECHANISM

The calcium bank is also called upon to perform its services during pregnancy, when a heavy drain may be placed on the mother's skeleton. If the expectant mother does not absorb a sufficient quantity of calcium from her daily food intake to satisfy her own bodily needs as well as the requirements of the developing fetus, the calcium bank is subject to heavy withdrawals. Then the skeleton of the pregnant woman becomes appreciably demineralized and may remain demineralized throughout her pregnancy. After the birth of the infant, when the burden of abnormal expenditures from the calcium bank is

removed and sufficient deposits can be made, the skeleton is restored
to normal mineralization. Of course, if the child is breastfed and the
calcium intake is still insufficient, the restoration of normal mineralization is delayed.

The process of demineralization concerns more than the simple
removal of calcium. In fact, the entire bone tissue is involved. When
calcium is withdrawn from the skeleton, the organic matrix immediately associated with the mineral is resorbed as well.

The effects of demineralization are often clearly recorded on x-ray
film. The x-rays of a person during a period of health, compared with
those taken during a heavy drain on the calcium reserves, show an
appreciable difference in density. In contrast to the well-mineralized
bone, the tissue of the demineralized bone is less dense; it is more
porous, and the individual spicules stand out in bold relief. When the

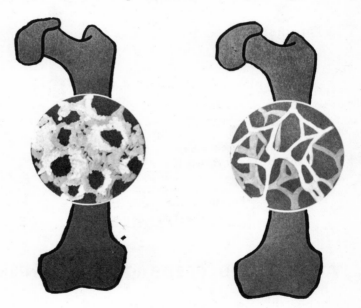

appropriate calcium balance is achieved, normal mineralization is
restored.

The mineral tissue affords the bone its rigidity and is an essential
component. A completely decalcified bone becomes flabby and bend-
able, like soft rubber. Of course, in the living human being, the calcium salts are never totally depleted, but in certain pathological
conditions the bones are so poorly mineralized that they approximate
a rubbery state. In rickets, for example, because of a deficiency of

vitamin D in the diet, mineralization is impaired. Since this vitamin is necessary for normal calcium metabolism and for the incorporation of calcium into the bones, children deficient in it often exhibit markedly deformed bones. In the absence of sufficient mineral, the bones of the legs gradually bend and become grotesquely bowed under the weight of the body. This happens, of course, over a period of time while the bones are growing. For this reason milk and other foods rich in calcium and vitamin D are important in the diet, not only for children but for adults as well.*

A third function of the skeleton is the housing of the blood-forming tissues of our bodies. During very early childhood, the red marrow, the blood-forming tissue, is found throughout the hollow spaces of the skeleton. Starting around 6 years of age, however, the marrow in the long bones of the arms and legs begins to lose its capacity to form blood cells. It changes into a yellow marrow and eventually, in old age, into a white fatty marrow. In adulthood, therefore, the blood-forming marrow is limited to the hollow spaces of the skull and face, the vertebrae, the ribs, and the pelvis.

In the event that a person is suffering from a disease of the blood, such as leukemia, the source of the trouble often may be located where blood cells are manufactured. For this reason, bone marrow samples are taken to determine whether the formative tissues are functioning properly.

The fourth and perhaps most remarkable function of the skeleton is its capacity to grow and to alter its form. We have previously discussed this function in following the skeleton through its lengthy journey of growth and maturation. The skeleton also plays a significant role in forensic medicine. In the next chapter we shall see how physical anthropologists have used the skeleton to carry on their work as bone detectives.

---

*A growing body of evidence supports the contention that vitamin D deficiency is not a dietary deficiency at all but is actually a hormonal inadequacy. This thesis states that sunlight produces the hormone calciferol in the deep layers of the skin. A significant lack of ultraviolet radiation (sunlight) results in a deficiency of this hormone and the appearance of rickets. The vitamin D_2 supplement is effective because it contains the necessary hormone and not because of its nutritional value. Cod liver oil is effective in combating rickets because fish, in contrast to birds and mammals, are able to produce calciferol without ultraviolet light.

Chapter 6

THE BONE DETECTIVE

Not infrequently newspapers report the discovery of a severely decomposed corpse in a secluded area. In extreme cases, only the skeleton is available to help determine whether death was due to natural causes or foul play. Lacking the usual articles of identification, the authorities may seek the assistance of a specialist in physical anthropology, sometimes called a bone detective. To identify the deceased person, three facts provide a starting point—race, sex, and age at death.

Age is assessed by determining skeletal age at death. In other words, the level of skeletal maturation is evaluated by studying the individual bones and observing their distinctive maturational features. The skull is examined to evaluate the degree of suture closure; the vertebrae and long bones are appraised to ascertain the extent of lipping; the stages of development of the various surfaces of the hip bones are analyzed; and the texture of the bone tissue is studied. From this type of examination, the bone detective is able to derive a skeletal age for each of the bones. By averaging the ages, he arrives at a composite age, which is regarded as the approximate chronological age at death. Of course, there may be an inherent error in this estimate because skeletal age depends upon the level of maturation of the skeleton, on whether the skeleton belonged to an average, rapid, or slow maturing individual. Therefore, the bone detective assumes that the person had developed at an average rate and then, for a margin of error, he calculates an estimated age at death (a skeletal age) with a year or two leeway on either side.

The next task is to determine the sex of the individual, a considerable challenge when only the skeleton or part of it is available for appraisal. The first consideration is size. Although the male skeleton is usually larger than the female, sheer size is an insufficient basis for reliable diagnosis of sex. Other factors must be considered. The male skeleton bears the marks of a heavier musculature and is usually more rugged and massive in appearance. It exhibits strongly developed sites of muscle attachment in the form of ridges and crests. Also, the male skeleton tends to be heavy, while the female skeleton is more gracile, more delicate. A tentative conclusion based on these differences can be reached, but if only these clues were used, incorrect judgments sometimes would be made because the features characteristic of the male are occasionally found in the female as well. A firmer basis for discrimination may be obtained by a detailed analysis of specific anatomical features.

The pelvis is the most reliable unit of the skeleton for determining sex. The female pelvis, peculiarly adapted to the birth process, differs in shape and proportions from the male. In the female, the birth canal is wide and oval-shaped, while in the male the inlet is small and heart-shaped. The angle formed by the right and left pubic bones and the angle of the sciatic notch tend to be much more acute in the male, and the sacrum tends to be narrower. From the pelvis alone, the bone detective can correctly assess the sex of the skeleton about 85 to 90 percent of the time.

Since the investigator is interested in obtaining the highest possible level of accuracy, he systematically examines all available regions of the skeleton. The skull is carefully evaluated in terms of a series of diagnostic sex traits. The supraorbital ridges in the male, as is evident in the illustration, are well-developed and contribute to a conspicuously sloping forehead. In contrast, the ridges are virtually

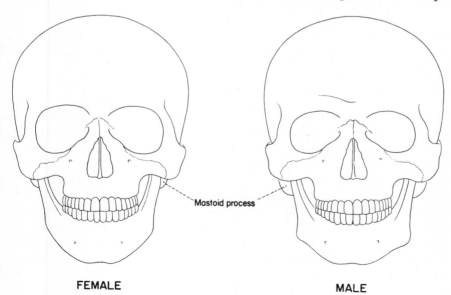

Mastoid process

FEMALE   MALE

absent in the female; she retains the infantile configuration of a globular forehead. In the male, the orbits, the holes for the eyeballs, tend to be square and small in relation to the size of the skull. The female, however, exhibits relatively large and rounded orbits.

The cheekbones of the male tend to be massive and flare outward, whereas in the female they tend to be lighter and more compressed. The mastoid processes, the nubbins of bone in back of the ears, are large and prominent in the male and small and delicate in the female. The palate in the male is broader and larger and tends to be U-shaped, while that of the female is small and parabolic. The occipital condyles, the processes on the bottom of the skull that attach the skull to the vertebral column, are large in the male and small in the female.

As a final example, from the total list of specific traits used to determine sex, we shall mention the external occipital protuberance (also called "inion"), the formal name for a nubbin of bone that projects downward at the back of the skull and is especially promi-

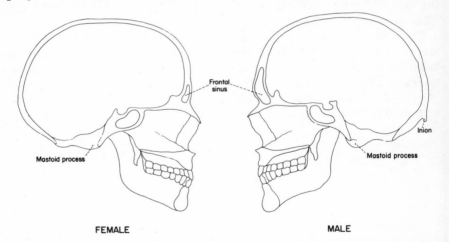

FEMALE   MALE

nent in males. Usually this bony projection is poorly developed in the female. By considering all the areas of the skeleton, a reliable judgment can be made in about 98 percent of the cases.

A third significant factor in identifying an unknown skeleton is race. The determination of race is based on a number of specific traits that are characteristic of each of the major groupings of mankind, the Caucasoid, the Negroid, and the Mongoloid. Although the entire skeleton is used by the bone detective in determining race, we shall use only the skull to describe some of the major differences in these three groups.

The Negro skull tends to be long and narrow and relatively low. The face is narrow and flat at the upper portion and exhibits a depressed nasal bridge. The bony opening of the nose is classically short and wide. The Negro skull is characterized by lower facial prognathism; that is, the jaws project forward, giving a concave appearance to the upper portion of the face.

The Mongoloid skull, in contrast, tends to be quite broad and round. The face is typically wide, with very prominent cheek bones and a flat nasal bridge. Classically, the Mongoloid face has a protrusion of the alveolus, the portion of the jaws that houses the teeth. The angle of the jaw is extremely wide and flares outward.

The Caucasoid skull varies considerably in shape, ranging from long and narrow in some skulls to short and broad in others. In contrast to the Negroid and Mongoloid faces, the Caucasoid face characteristically exhibits a straight profile; neither the jaws nor the portion of the bones holding the teeth protrudes. In fact, in the Caucasoid, it is the nasal area of the face which is most prominent. The bony part of the nose tends to be long and narrow and quite protrusive, with a high nasal bridge.

Perhaps the task of determining race from an examination of the skull seems routine and simple. Actually, it is a complex problem, because the traits typical of each race are found in varying degrees in the other races. Racial interbreeding has further complicated the problems of identification.

Using the information that the skeleton affords concerning the age, sex, and race of the deceased, investigators have actually attempted to reconstruct the individual's face as it appeared in life. In order to accomplish this task, it is necessary to determine the thickness of the soft tissues overlying the different parts of the facial skeleton. To obtain this data, anatomists since the 1880's have used cadavers. Pins are stuck into the skin and flesh of the face at many different areas until they contact the underlying bone. In this way, it is possible to determine the total thickness of the skin, fat, and muscle in different parts of the face. By studying hundreds of individuals, the anatomist has been able to build up an average picture of the thickness of the soft tissue of the head and face. Using this information, he artistically reconstructs the facial features of the deceased individual from the skull alone. This is accomplished by placing blocks of clay of appropriate thickness at each of the key sites on the face of the skeleton. The blocks are then connected by strips of clay until the entire face is modeled, as shown in the accompanying illustration. Naturally, the reconstructed face is based on studies of average faces. If the deceased individual differed markedly from the average in body weight, for example, the reconstructed face would deviate considerably from the actual face of the dead person.

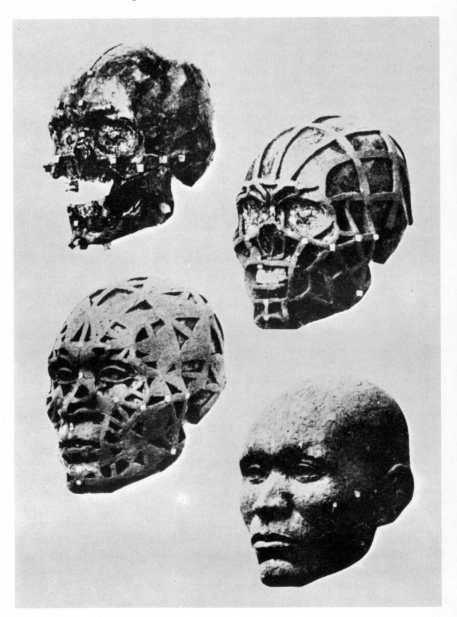

Another factor that may help to identify the deceased person is stature. Even if only portions of the skeleton are available, this measurement may be obtained with a reasonable degree of accuracy. Minimally, the bone detective needs but a single long bone from an arm or leg to make his estimate. Anatomists and physical anthropologists have studied the relative contributions of the long bones to total stature in many thousands of individuals of both sexes in all three major racial stocks. From these studies, they have established a series of formulas for computing an estimated stature of the individual, using only the length of one of the long bones. The estimated height contains a margin of error of approximately an inch or two in either direction.

The teeth also serve as a valuable adjunct for identification. In technological societies such as ours, most persons exhibit considerable dental repair. The pattern of fillings, inlays, and crowns for each person is nearly as unique as his fingerprints. Dental records, therefore, are checked to provide a specific basis for identification.

The investigator also examines the site in which the remains were found. The climatic conditions of the region are of considerable importance in resolving the circumstances of death. From the state of preservation of the body, an estimate can be made of the time that has elapsed since death. Preservation varies according to the climate and the type of soil in which the body is interred. In a tropical climate, decay takes place rapidly and even the skeleton soon begins to decompose and crumble, while in an arid or desert climate, preservation is excellent and decay takes place much more slowly. You may have read accounts of archeologists, working in the American southwest and in Mexico, discovering the remains of naturally mummified Indians who had died hundreds of years ago. Because of the aridity of the climate and the excellent preservative nature of the sandy soil, decomposition was extremely retarded, so that even the skin, hair, and dried muscle tissue remained intact to a remarkable degree.

Now that we have uncovered the methods through which the bone detective obtains his evidence, let us see how he applies his science in real-life mysteries. The following cases are based on investigations conducted by Dr. Wilton M. Krogman.

The Skeleton in the Culvert

In 1935, a partially buried skeleton was discovered by a group of children playing in a culvert. The entire skeleton was exhumed by the police, taken to the county morgue, and turned over to Dr. Krogman for determination of age, sex, race, and other data helpful in identifying the person. Establishing the sex of the individual was a relatively easy matter. The skeleton was found to be small; the long bones were slender and delicate and the bone surfaces and ridges for muscle attachment were weakly developed. These features are characteristic of the female skeleton. Examination of specific sex-related traits corroborated this initial impression. The pelvis exhibited a relatively large oval inlet, a broad subpubic angle and an open sciatic notch—all female traits. By arranging the bones in their proper relationship, it was possible to arrive at a fairly accurate estimate of the woman's stature, which was found to be 5 feet 4 inches.

The texture of the bones and the degree of lipping offered additional information concerning age at death. The texture of all the bones was characteristic of a person 35 years of age, while the absence of lipping indicated that the person had not yet reached 40. Turning to the skull, since the sutures of the cranial vault were all closed, an age of at least 35 years was assumed. Summing up all the criteria, an age range of from 28 to 40 years was revealed, with the majority of traits suggesting an age of 35.

The determination of race or stock presented a complex problem. Decision was made on the basis of assessing 14 traits of the skull and 21 measurements of the skeleton. Evaluation of this data indicated racial admixture; it was concluded that the subject was part Negro and part Caucasian (of Southern European origin). The precise degree of racial mixture could not be determined, but the white component appeared to dominate.

The evidence of elapsed time since burial of the body was systematically evaluated. A conclusion was based on the analysis of the soil content, its state of acidity, the water drainage in the area, the nature of the interment, and the state of decomposition of the remains. Strands of muscle tissue and ligaments still adhered to some of the bones. Further, the periosteum (the surrounding membrane) was still intact on many of the bones. All the long bones retained a considerable amount of marrow substance as well as the organic matrix of the bone tissue. Bits of brain tissue adhered to some of the skull bones. The texture of the bone tissue was not dry, light, and brittle, as is characteristic of the archeological relic. Since the soil in the culvert and the soil on the skeleton matched exactly, it was assumed that the culvert was the original grave. It was also inferred that the body had been placed on a natural ledge in the culvert and had been hastily and superficially covered. Collating all the available data led to the conclusion that the body had been interred approximately a year prior to its discovery. If the body had been interred for an appreciably longer period of time, all the organic material would have been decayed.

In summary, then, Dr. Krogman's analysis led him to conclude that the remains belonged to a woman, approximately 35 years old, 5 feet 4 inches tall, of mixed ancestry (part Negro and part Southern European Caucasian), who was killed and hastily buried about a year before her remains were discovered. The skeleton itself provided no data relating to the cause of death.

The Runaway Millionaire

In the early 1920's, R. J., a youth of mixed Indian and Negro ancestry disappeared from his home on a Seminole reservation in Oklahoma. Several weeks later, in Blue Mountain, Arkansas, a boy answering the general description of the missing R. J. was killed while attempting to board a freight train. A coroner's inquest issued a verdict of accidental death, and the boy was buried on the right-of-way of the railroad. Subsequently, oil was discovered on land that had been allotted to the boy because of his status as an Indian. His father was appointed administrator of the estate. The title to the land was contested by an oil company that held interests in the area. It became necessary, therefore, to establish the fact of R. J.'s death and his father's legal right to inheritance.

The corpse was exhumed and examined by an anatomist retained by the oil company. It was his considered judgment that the remains were of an adult male of at least 30 years of age at the time of death and not those of the missing boy. The attorney for the father questioned this conclusion and requested a re-examination, which was conducted by Dr. W. M. Krogman.

To resolve the primary issue of age at death, the following skeletal features were evaluated.

The Skull: All the sutures of the cranial vault were found to be open both internally and externally. The bones at the base of the skull, however, were firmly united. Since the sutures in the cranial vault begin to fuse in the early twenties, the subject was probably under 25 years of age. Fusion of the bones at the base of the skull, however, is usually completed before 20. These facts, therefore, tend to point to an estimated age at death of from 16 to 25 years.

All the permanent teeth, including the third molars, were erupted and in functional occlusion, yet the molars evidenced no signs of wear. There were no filled or decayed teeth. The third molars usually erupt between 15 and 18 years of age; the fact that they showed no signs of wear strongly suggested relatively recent eruption. The absence of caries also implied a fairly young person.

The Scapulae: The processes of the scapulae (shoulder blades) were united. The joints of the scapulae, at their attachments with the upper arms, had completely developed surfaces without rim formation.

The Clavicles: The epiphyses at the ends of the clavicles (collarbones) were not united on either the right or left side.

The Long Bones: In both humeri (upper arm bones), the epiphyses at the elbows were united, while those at the shoulders were not. In both bones of the lower arm (radius and ulna), the epiphyses at the elbow joints were united, while those of the wrists were fully open. In the thighbones, the epiphyses at the hips were completely united, while those at the knee joints were only partially united. The heads of the thighbones at the hip joints, however, were not united to their shafts. In both bones of the lower leg, the epiphyses at the knees were in the process of uniting but had not yet closed; the epiphyses at the ankles had recently united. None of the articular surfaces of the long bones showed signs of lipping.

The Pelvis, The Ribs, The Vertebrae: All evidenced epiphyses that were not united. The pubic symphysis of the pelvis (the junction of the right and left halves of the pelvis at the front of the body) showed traces of billowing of the surfaces compatible with an age of 18 to 20 years.

The following table lists the epiphyses studied by Dr. Krogman, the time of expected union in the male, and the condition of each of the epiphyses for the deceased. In the table, a plus indicates that fusion had occurred; a minus designates that the epiphysis was still open; a combined plus and minus for a given epiphysis means that union had begun but was not yet completed.

The cumulative evidence in terms of gross examination of the skeleton, detailed x-ray analysis, and systematic evaluation of each of the epiphyses, as indicated in the table, established a composite age of at least 18 and not more than 19 years at death.

ASSESSMENT OF EPIPHYSEAL UNION

Epiphysis	Expected Union (yr.)	Degree of Fusion in Remains Studied
Scapula		
Acromial process	18.0–19.0	+
Coracoid process	16.0–17.0	+
Vertebral margin	20.0–21.0	− (?)
Clavicle		
Sternal end	25.0–28.0	−
Acromial end	19.0–20.0	?
Humerus		
Head	19.5–20.5	−
Distal	14.0–15.0	+
Medial epicondyle	15.0–16.0	+
Radius		
Proximal	14.5–15.5	+
Distal	18.0–19.0	−
Ulna		
Proximal	14.5–15.5	+
Distal	18.0–19.0	−
Pelvis		
Primary elements	14.0–15.0	+
Crest	18.0–19.0	−
Ischial tuberosity	19.0–20.0	−
Femur		
Head	17.0–18.0	−
Greater trochanter	17.0–18.0	+
Lesser trochanter	17.0–18.0	+
Distal	17.5–18.5	±
Tibia		
Proximal	17.5–18.5	±
Distal	15.5–16.5	+
Fibula		
Proximal	17.5–18.5	±
Distal	15.5–16.5	+

The problem of racial background was resolved by assessing a series of discrete structural features that characterize each of the major groupings of mankind and by a series of extensive anthropometric measurements based on the study of the different races. After evaluating this material, it was concluded that the remains were those of a person of mixed racial ancestry exhibiting both American Indian and Negro traits.

Thus, it was established that the remains, in fact, were those of R. J., and a verdict was rendered awarding R. J.'s father a clear title to the land.

Chapter 7

THE NATURE OF WEIGHT

Our culture is a weight-conscious culture. Both the cultural ideal and the well-advertised medical admonitions against overweight persistently thrust our physiques into awareness. *En masse,* men and women alike have become weight watchers.

In the minds of most people, the slogan "Watch your weight" is equivalent to "Watch your fat!" But this is much too narrow a view of body weight, since the body is an aggregate of many tissues. It includes muscle, bone, gut, brain, nerves, blood, skin; even the toe nails, finger nails, and hair contribute to body weight.

When we step on a scale we obtain a single gross measurement of our total body mass, but this measurement is deceptive. People with identical weights often possess markedly dissimilar physiques. To understand the real nature of weight we must determine the relative contribution of the individual organs and tissues of the body to the total weight of the growing person.

Unfortunately, it is not possible to take the living individual apart and weigh each of his organs and tissues separately. How, then, may these data be obtained? The early anatomists were able to determine the contribution of the individual tissues to total body weight by dissecting and studying hundreds of cadavers of fetuses, newborn infants, children and adults. These studies revealed a fundamental difference in the pattern of growth of the individual organs *before* and *after* birth. During the fetal period, the major organs of the body exhibit the same basic pattern of growth. The brain, stomach, kidneys, lungs, heart, uterus, and adrenal glands all increase in weight at approximately the same rate. Consequently, the percentile curves for individual organs and for total body weight, as well, reflect the uniformity of prenatal growth.

In the first year after birth a striking dispersion occurs. One after the other, these organs depart from the prenatal growth pattern. The adrenals and uterus actually decrease in weight, while the heart,

109

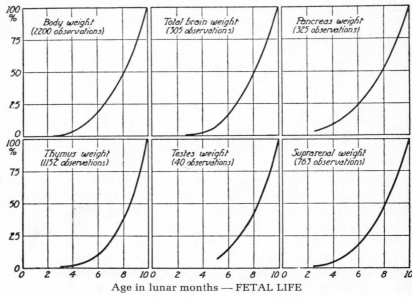

Age in lunar months — FETAL LIFE

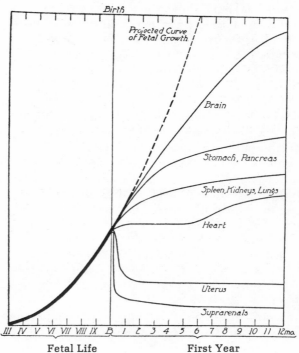

Fetal Life   First Year

kidneys, lungs, and stomach progressively slow down in their rate of growth. The brain deviates the least from the fetal pattern. Thus, the first year of infancy represents a period of transition when basic organs and structures of the body establish their individual patterns of growth. The marked contrast between the first year of life and the prenatal period is shown in the preceding illustration.

The divergent patterns of growth established in the first year of postnatal life are reflected in the changing contribution of the different organs to total body weight at different ages. While some organs are significantly reduced in their percentage of total body weight, others are greatly increased. This shift can be illustrated by comparing the relative size of the brain and internal organs in the newborn

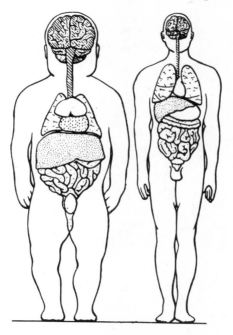

with those of an adult. The newborn infant possesses a disproportionately large brain and gut. At birth, the gut comprises 16 percent of total body weight, while in the adult these organs contribute only 11 percent. The brain and spinal chord are even more dramatically reduced during development, contributing 15 percent to the total weight of the newborn and only 3 percent to that of the adult. In contrast, muscle mass, which comprises 25 percent of body weight in the newborn, increases to 43 percent in the adult.

The study of cadavers provided a general picture of the changing contribution of the individual tissues to total body weight. In the early 1940's, new techniques were devised making possible the detailed study of tissues and organs in the living person. These techniques utilize x-rays to penetrate the body and make the internal organs visible. On the x-ray film, for example, fat, muscle, bone, and skin can be discerned, as illustrated in the picture of the leg of an 11-year-old boy.

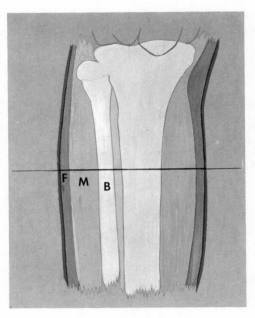

Fat and skin	= 17 mm
Muscle	= 50 mm
Bone	= 30 mm
Total calf width	= 97 mm

The relative contribution of each of the tissues is determined by measurements taken at the widest diameter of the calf. A line is drawn at this level on the x-ray film. In the example, the width of the fat and skin is 17 mm; muscle width is 50 mm; and bone width is 30 mm. Thus, total width of the calf is 97 mm. By repeating this procedure at intervals from birth to adulthood, it is possible to assess the changing contributions of each of the major tissues to total calf breadth. In addition, sex differences in the rate of growth for each of the three tissues can be studied.

Using the procedure described above, one study analyzed the growth of the tissues of the leg in males and females from birth through 30 years of age. The pattern of growth for total calf breadth was found to be essentially the same in both sexes, as shown in the upper curves of the illustration. If this pair of curves comprised the

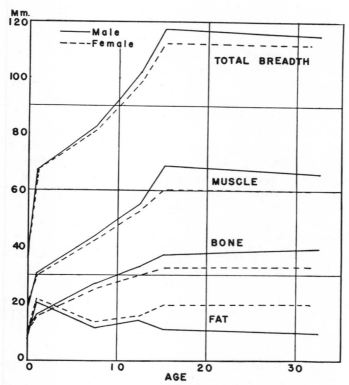

total available data, one might assume that the male and the female leg grow in the same way and are composed of the same relative volume of tissues. We know, however, that the male and female leg are distinctly different, and the curves of growth for the individual tissues prove it.

Although the pattern and amount of growth of the individual tissues are quite similar for boys and girls during childhood, at adolescence the sexes part company. The male exhibits a great burst of muscle and bone growth, while the female has the definite edge in growth of fat.

An examination of the curves of growth for fat provides an instructive picture of changes occurring at different periods of development. At 1 year of age, the breadth of fat in the calf is as great, in both sexes, as it is at any point in the life span. In fact, in the male the breadth of fat at 1 year is twice that of adulthood, while in the female it is equal to that of adulthood. After the first year, fat decreases markedly in both sexes, and the period of chubby babyhood comes to an end. The sexes are comparable until the approach of adolescence. At this point, the female exhibits a significant increase in breadth of fat, while the male undergoes a further loss.

Although we have only considered the tissue differences along a single diameter of the calf, what is happening in the leg is reflected in other parts of the body. In other words, in the male, muscle and bone growth are dominant throughout the body, while fat plays a more significant role in the female.

While it may seem odd to be measuring the width of tissues on x-ray film, this is actually a widely used method of estimating relative weight contributions in the living person. In applying the technique, it is assumed that a relationship exists between surface area of tissues and their relative contributions to total body weight. Although such a relationship exists, since the specific gravity of the tissues is not the same, two tissues covering the same area on a film can contribute unequally to weight.

Thus far we have examined one technique for assessing differential tissue growth in the calf. A somewhat different approach has been devised for measuring the *total* surface area of each of the tissues. In this technique, the x-ray picture is literally cut up. To illustrate, we shall return to the x-ray film of the 11-year-old boy discussed earlier. First, the background film surrounding the shadow of the leg is cut away. Next, the portions of the x-ray representing the bones at the knee and ankle are cut off. Then the strips of film representing the fat and skin (F) on the inside and outside of the leg are removed. Finally, the strips of muscle (M) are separated from the bone (B). We now have seven irregular strips of x-ray film, two for fat and skin, three for muscle, and two for bone. The strips of film representing each tissue are then weighed. For our 11-year-old boy, we find that the fat and skin weigh 1.54 grams, the muscle weighs 2.34 grams, and the bone weighs 2.54 grams. The combined weight of the three tissues is 6.42 grams. It must be remembered, of course, that we are only weighing x-ray film *not* fat, muscle, and bone; nevertheless, the relative weights of the strips of film provide information concerning the

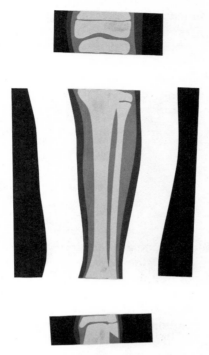

relative surface area contributed by each of the tissues. This sug-gests, in turn, the approximate contributions of each of the tissues to total body weight.

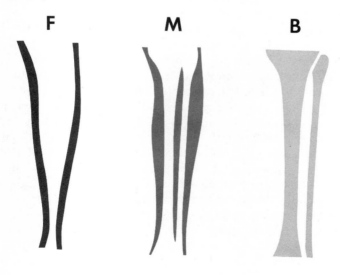

How do the results of the two methods compare? The table shows the percent contributed by each tissue according to the surface area method and the widest calf diameter method. Clearly, the results are not identical. The reason is that in weighing total surface area the

Comparison of Two Methods of Assessing the Tissues of the Leg

Tissues	Surface Area Method	Widest Calf Diameter Method
Bone	39.6%	30.9%
Muscle	36.4%	51.6%
Fat and Skin	24.0%	17.5%
Total Calf	100%	100%

large, heavy nubbin of bone at the top of the calf is included, but in the calf diameter method, since measures are taken across the middle of the leg, this large nubbin is excluded. Of the three tissues, fat shows the greatest similarity when the two methods are compared. The data show considerable difference in bone and muscle contributions; however, when bone and muscle are combined, whether total calf breadth or total surface area is used, studies report fairly high correlations between assessments taken from x-ray films and actual body weights.

We have emphasized that total body weight is an aggregate of many tissues and that each of the tissues may contribute differentially in determining the weight of the individual. Keeping these facts in mind, we shall now discuss the changes occurring in total body weight from birth through adulthood.

Immediately after birth both boys and girls evidence a sharp loss of weight. This loss is due to the traumatic effects of birth itself and the demands it puts on the newborn infant. Consider for a moment the impact of being suddenly thrust out of a warm, stable uterine environment, where nourishment is provided automatically, to an outside world where one must eat, breathe, experience temperature shifts, and respond to a multiplicity of stimuli. The healthy infant soon adapts to the new environment and recovers the lost weight. During the first few months he exhibits dramatic weight increases and acquires the chubbiness of infancy.

The period of chubbiness lasts until the age of 2 or 2½, when the baby fat is lost. At the completion of infancy, much to the consternation of parents, the child's appetite wanes and he becomes slender. From this time until about 10 years of age in girls, and somewhat later in boys, weight shows a slow but steady increase. Then at puberty weight spurts dramatically. Boys show an appreciable gain in muscle mass, while girls begin to acquire subcutaneous fatty tissue, anticipating the rounded contours of the adult female. Approaching young adulthood, the rate of gain in weight slows down in both sexes. Subsequent weight gain reflects not so much the growth factor but rather individual physique and gastronomical disposition. The changing trends and the shifts in weight gain from birth to young adulthood are shown in the accompanying curves for boys and girls.

The pattern of growth in weight is similar to the pattern of growth in height. From birth to young adulthood, for both sexes, the curve of growth in weight exhibits three phases, two decelerating and one

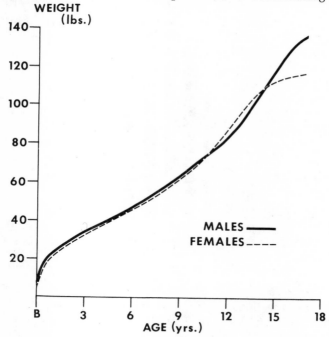

accelerating. And again, like stature, differences between the sexes are due to earlier attainment in girls and greater attainment in boys.

Although the above growth curves are based on a population of Caucasian children born and reared in Iowa, they are not unique to this one segment of the population; they are representative of all human beings. Differences exist between the races in the *timing* of the phases of growth and the *magnitude* or amount of actual weight increase, but the basic pattern of growth remains the same.

Although the average curves for height and weight show a similarity in pattern, certain fundamental differences may be discerned. One of these differences is shown by comparing the heights and weights of a group of 10-year-old boys. At a glance it is clear that the distributions of the heights and weights in this population are quite dissimilar in shape. The statures are distributed in a bell-shaped and symmetrical form, while the weight distribution is lopsided or asymmetrical. In the distribution for statures, it is apparent that one bar

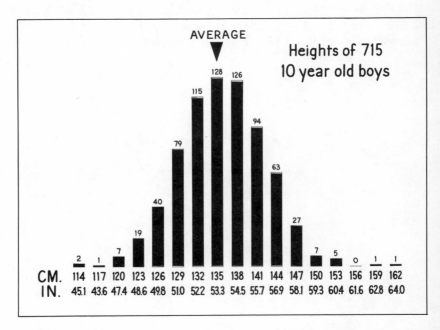

is higher than the others. This bar represents the height that occurs most frequently in the sample of 715 boys. It indicates that 128 of the total group are 53 inches tall. The bars to the left of the peak group represent successively shorter height groupings; the bars to the right of the peak group represent successively taller height groupings. As we move away from the peak group, the bars become progressively shorter because there are fewer boys in each group. The tallest group contains but one boy—64 inches in height; the shortest has two boys in it—45 inches tall. By connecting the tops of all the bars, it is possible to construct a bell-shaped curve. This curve is called a *frequency distribution* for the obvious reason that it shows the frequency of each of the heights in this population of boys. The average height is obtained by adding all the heights and dividing by the total number of boys in the sample. You will note that the average (53 inches) falls at the center of the peak group. This is because the frequency distribution is symmetrical.

The frequency distribution of weight for the same population presents a different picture. When the tops of the bars of the different weight groupings are connected, the result is a lopsided or asymmetrical curve. This result is obtained because there are more heavy

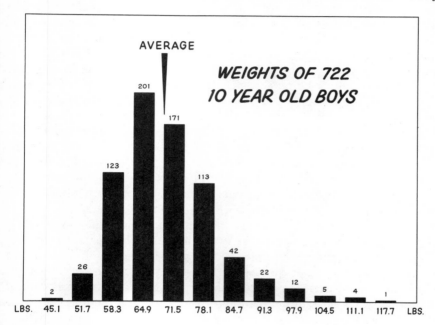

groups of boys to the right of the peak group than there are light groups of boys to the left of it. The unbalanced distribution is not a quirk of 10-year-old boys; it is representative of weight at any chronological age.

Since the distribution for weight is asymmetrical, the average weight does not fall at the peak group. This is demonstrated in the above illustration in which the peak group represents a weight of 64.9 pounds, whereas the average weight of the population is 69.9 pounds. The long tail of heavy weights is pulling the average to the right, away from the peak group. Therefore, whenever a trait is distributed asymmetrically, the average is not a good indicator of the center of the distribution. In the above illustration, for example, 409 boys are lighter than the average while only 313 boys are heavier.

An even more glaring example of the undue influence exerted by an extreme score can be demonstrated by computing the average income of a sample of ten men. If nine men each earn $5,000 annually and the tenth earns $100,000, the average income of the group is $14,500. How representative is this average? It exceeds 90 percent of the incomes by nearly three times, while it falls short of the remaining 10 percent by $85,500! Obviously, given an asymmetrical distribution of this kind, the average is a deceptive measure of central tendency. A more valid measure is gained by determining the *median* or middle score. To obtain the median, each individual is given a score or percentile rank so that the fiftieth percentile, the middle individual in the distribution, can be located. In the case of our ten incomes, ranking them from lowest to highest, we find that the middle income, represented by the fifth and sixth men, is $5,000. Our fiftieth percentile income actually describes 90 percent of this population and thus is a better measure of central tendency than the average.

Looking again at the weight distribution, we find that the fiftieth percentile weight is 68.5 pounds, or 1.4 pounds less than the average. The fiftieth percentile weight was obtained by counting 361 boys from the left, or half the number of boys in the distribution. In working with weight distributions, human developmentalists use the fiftieth percentile (the median) in preference to the average.

Earlier, we stated that at any chronological age, the distribution for weight is asymmetrical. Why is this so? The reason probably stems from the intrinsic nature of weight as it relates to health and survival. Although a person can be extremely heavy and still survive, there is a lower limit to weight below which the individual cannot maintain life. This can be demonstrated by examining once more the distribuiton of weights for the 10-year-old boys. The median or fiftieth percentile weight for this population is 68.5 pounds. The heaviest boy weights 117.7 pounds, or 49.2 pounds more than the median boy. If he were matched by a comparably light boy, this individual would have to be 49.2 pounds lighter than the median. He would weigh only 19.3 pounds, less than the median weight of a 7-month-old infant! In contrast, a person can be either very tall or very short and still survive. In other words, in stature an equal number of people *can* be represented at the extremes of the distribution.

Weight exhibits a second important difference from stature. *Weight is reversible.* It is this peculiar characteristic that makes dieting and reducing possible. The reversibility of weight is demonstrated by the before and after pictures of extremely heavy people who have become slender on crash diets. Many magazines show the fat lady transformed into a sleek, attractive woman consistent with the cultural ideal. What the ads fail to say is that no matter how extensive the weight reduction, if the potential for a slim shape is not there, it cannot be achieved. The woman who is rawboned and heavily muscled cannot transform herself into a willowy size 10. Although fat can be largely eliminated without impairing the person's health, muscle and bone cannot be so easily reduced; the structure of the body cannot be changed. Shedding generous cushions of fat may only succeed in exposing a stocky, muscular physique.

Another consequence of extensive weight reduction occurs among middle- and old-aged dieters. In these persons, the skin, having lost the resiliency of youth, hangs loose in folds and appears to be several sizes too large for the body. Unfortunately, this skin cannot be shed.

When an individual is dissatisfied with the results of his diet, he may literally starve himself into a state of malnutrition, losing weight at the expense of good health. Although this may seem absurd, it is a course of action followed by too many people.

What is actually lost through human starvation? This question was investigated by a group of biochemists and nutritionists in their studies of volunteer conscientious objectors during World War II. In

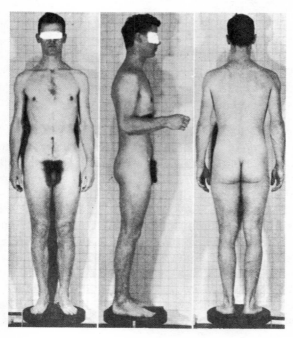

the first phase of the research, a group of healthy young men subsisted on an extremely impoverished diet for 24 weeks. During this
period of semistarvation, the men gradually lost their robust physiques and became severely emaciated. The illustration shows one
of the men at the beginning of the study and at the peak of semistarvation. The investigators were able to determine the relative

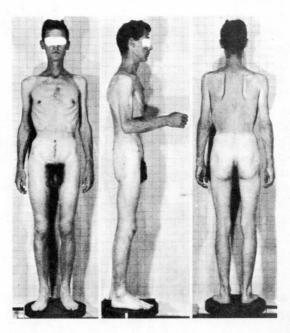

weight loss of many tissues of the body during human starvation.
Although the men lost about 24 percent of their total body weight,
the tissues of the body did not suffer equally. Some were more resistant to starvation than others. Bone mineral, for example, showed
almost no reduction, while fat, at the other extreme, was reduced by
69 percent of its original volume. A considerable amount of muscle
was also lost (41 percent); the gut and blood volume each lost 9 percent, while the brain was reduced only 4 percent. In extreme dieting,
then, the fat and muscle components are most severely affected,
while the brain and the bones are least susceptible to nutritional
deprivation.

The researchers did not leave their subjects in an emaciated state.
In the second phase of the study, the men were placed on optimal
diets in order to restore their body tissues and recover their original
physiques.

Clearly, a starvation diet is not the most desirable way to lose weight. The food we eat provides the building materials for the tissues of the body and supplies the fuel necessary for bodily activity. Excess fuel intake is not spilled off like an overflowing gas tank but is stored in the form of fat deposited between the muscle fibers, in the bodily spaces, and under the skin. Fuel intake is expressed in terms of calories, or units of heat production. If caloric intake is in excess of bodily needs, we are confronted with a problem of excess weight. This problem cannot be resolved by increased exercise alone. Even if a person walked 40 miles a day, he would burn up only 3600 calories, or the equivalent of 1 pound of fat. The best solution to the problem of excess weight is a plan that combines a balanced diet with appropriate exercise.

In our discussion of dieting, we have concentrated on the reversibility of body weight in adults. Weight reversibility is also an important concept in understanding the developing child. Any dietary plan must recognize two factors: the child requires enough nutrients to *maintain* the health of the tissues and bodily metabolism on the one hand and sufficient minerals and proteins *to enhance* bodily growth on the other. Since the child is a growing organism, we expect his weight to increase throughout the developmental period. Other than in the obese child on a special diet, an appreciable loss of weight is a matter of concern. It is often the first sign of systemic disturbance. Frequently the loss is due to physical illness. If the illness is of short duration, the child will soon recover the loss and resume his pattern of gain in weight. When the weight loss persists without apparent cause, an intensive pediatric examination is indicated. In the event that no organic basis for weight failure is revealed, the problem must be considered in terms of emotional disturbance. Referrals to child psychologists are sometimes made by parents and physicians who are alarmed by a child's continued loss of weight.

The psychologist may determine that the child is under stress due to a temporary crisis. For example, a new baby coming into the home may be sufficiently disturbing to cause an older sibling to stop eating and lose weight until he is reassured of his place of esteem and affection in the family. Any crisis, such as death or temporary absence of the mother or father from the home, can initiate a period of stress in the child that results in significant weight problems. With special support and affirmation, these crises are usually weathered and the child soon is restored to health. More deep-seated problems require prolonged treatment and often intensive psychotherapy. In *anorexia nervosa*, the impairment is so severe that eating becomes a situation of extreme threat; while undergoing psychiatric treatment the child may even require intravenous feeding.

Significant weight reversal is an obvious source of concern to parents, but a relative weight loss can also create anxiety. By relative weight loss, we mean that the child is not adding weight proportionate to his increase in height. There are times during development when the child is typically thinning out, extremely active in his behavior, and uninterested in eating. This is particularly characteristic of the phase of development beginning at about 3 years of age and continuing through the age of 7 or 8. At this time parents often seek the support of their physician; they may even seek reassurance from the height-weight tables printed on penny scales. They want to know if their child's weight is normal for his height.

How much should a child weigh for his height? How can we determine whether a child is "normal"? To answer these questions, we must examine the nature of height-weight relationships.

Chapter 8

SO WHO'S NORMAL?

The subject of acceptable height-weight relationships has probably received more concentrated attention than any other aspect of physical development. Before the advent of our computer society, the individual trusted his own subjective assessment of his body build and his degree of overweight. Today, however, he looks for a scientific definition of obesity; his estimate is subordinated to standards derived from statistical analysis.

In this chapter, we shall examine standards for determining acceptable weight limits for the individual. We shall describe some of the major innovations resulting in the improvement of techniques for assessing and evaluating height-weight relationships.

Fascination with the Average

What constitutes ideal weight? To answer this question, the early investigators fastened onto the idea that weight bears a meaningful relationship to stature and that stature could be used as a basis for determining "correct" weight. Consequently, different populations were surveyed in order to determine the average weight for height at each age. These data, arranged in tabular form, constituted a set of norms and soon these norms became the standard for *normal*. Thus, the statistical average became synonymous with normal or ideal height-weight relationship. Given a person's age and height, one could quickly determine his "correct" weight. "How much should I weigh for my height?" became the standard question, and the answer could be purchased for a penny at the corner scale! The penny scale communicated the idea that there was a single, most desirable weight for a given height.

Let us examine these height-weight tables to see how they were devised and the premises on which they were based. One of the earliest of these tables, published in 1923, is reproduced below.

BALDWIN - WOOD

WEIGHT - HEIGHT - AGE TABLE FOR BOYS OF SCHOOL AGE

Height Inches	5	6	7	8	9	10	11	12	13	14	15	16	17	18	19
38	34	34													
39	35	35													
40	36	36													
41	38	38	38												
42	39	39	39	39											
43	41	41	41	41											
44	44	44	44	44											
45	46	46	46	46	46										
46	47	48	48	48	48										
47	49	50	50	50	50	50									
48		52	53	53	53	53									
49		55	55	55	55	55	55								
50		57	58	58	58	58	58	58							
51			**61**	**61**	**61**	**61**	**61**	**61**							
52			63	64	64	64	64	64	64						
53			66	67	67	67	67	68	68						
54				70	70	70	70	71	71	72					
55				72	72	73	73	74	74	74					
56				75	76	77	77	77	78	**78**	80				
57					79	80	81	81	82	83	83				
58					83	84	84	85	85	86	87				
59						87	88	89	89	90	90	90			
60						91	92	92	93	94	95	96			
61							95	96	97	99	100	103	106		
62							100	101	102	103	104	107	111	116	
63							105	106	107	108	110	113	118	123	127
64															

Assume that our subject is a 10-year-old boy who is 51 inches tall. The table indicates that the average weight for 10-year-old boys of this height is 61 pounds. In fact, according to the table, this is the average weight for any boy of that height between the ages of 7 and 12. Although the table does not categorically say that a boy 51 inches tall *must* weigh 61 pounds, in the absence of qualifying data, parents often interpret the table this way and become alarmed if their child deviates appreciably from the indicated weight. One may ask, how characteristic is the listed weight of 61 pounds for 10-year-old boys who are 51 inches tall? How many 10-year-olds approximate this weight? The answers to these questions can be obtained by examining the distribution of heights and weights in a healthy group of 10-year-old boys. Looking first at the distribution of heights for a sample of 715 boys, only 79 are found to be 51 inches tall. The

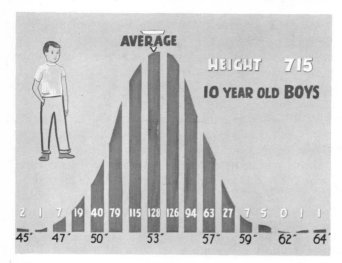

next graph shows that the 79 boys range in weight from 53 pounds for the lightest to 73 pounds for the heaviest.

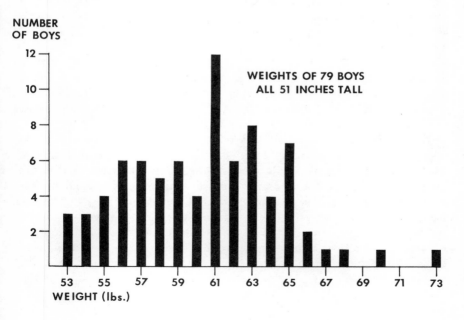

How many of the 79 boys actually weigh 61 pounds, the weight listed in the table as the "average" for 10-year-olds? Just 12 of the boys match this weight; that is, only 15 percent have the same weight as that listed in the Baldwin-Wood tables. The remaining 67 boys are either lighter or heavier. If we regard the average as an index of normalcy, then 67 boys in the group must be regarded as abnormal. It is evident that the old-time height-weight tables do not consider the tremendous individual variation in weight that occurs at any given height in a normal population of children. Consequently, use of these tables to determine "correct" weight will be misleading.

The average describes the population as a totality; it describes everyone in general and no one in particular. It provides no insight into the individuals who make up the population and the manner in which they are distributed above and below the average. Further, the Baldwin-Wood tables are a static device since they assess a child's *status* at one point in time. An effective appraisal of height and weight should also evaluate a child's *progress* through time.

Recognizing Individual Variation

An important modification in the development of height-weight charts occurred when these devices incorporated guidelines that recognized individual variation and facilitated the study of a child's progress through time. Such a graphic method based on the analysis of 3,000 Iowa children of both sexes was devised by the Department of Pediatrics at the State University of Iowa in the 1940's. The two charts for boys depicted here cover the age span from birth through 18 years. The first chart refers to the period from birth to 6 years, while the second overlaps the first and covers the age range from 5 to 18 years. Let us examine these charts to determine their use in assessing a child's status and his progress.

Notice that there are two sets of curves, the upper pertaining to weight and the lower related to height. Chronological age is indicated on the horizontal axis, while weight and height are recorded on the vertical axis. In order to use the graph properly, one must understand how the curves were constructed and what they represent. Looking first at height, the center curve represents the average height in inches at each chronological age. The upper and lower curves represent deviations from the average. These deviations are based on a statistical concept called the *standard deviation*. Thus, the upper and lower curves demarcate the limits of one standard devia-

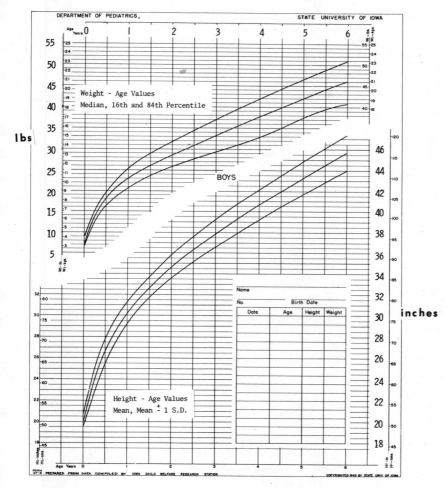

tion, plus and minus. Within these limits will be found *the middle two-thirds*, or more precisely *the middle* 68 percent of the population at each chronological age. In other words, the standard deviation is a measure of dispersion. It tells us how the population is spread out above and below the average. It indicates the range of variation expressed by the middle 68 percent of the population. Put more simply, two out of every three boys at each chronological age will fall between the upper and lower lines. The remaining 32 percent at each age will be divided equally above and below the curves representing plus and minus one standard deviation. Thus,

16 boys out of every 100 will be taller than the height indicated by
the plus one standard deviation line and 16 will be shorter than the
points along the line representing minus one standard deviation.

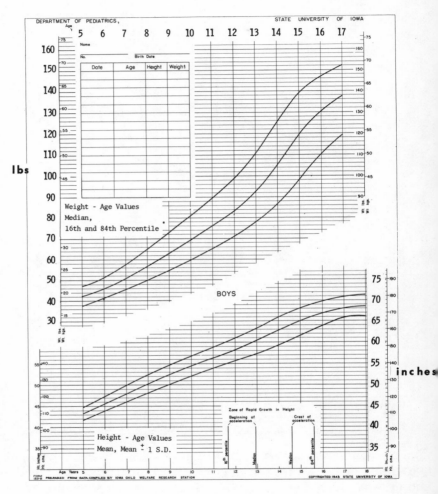

The three curves for stature depicted on the Iowa graph are
actually based on the frequency distribution for stature constructed
at each chronological age. The reason that only 16 boys out of
every 100 fall above or below plus and minus one standard deviation is the very shape of the frequency distribution. As we move
away from the average, fewer and fewer boys are represented.
In each sample of normal boys there are relatively few extremely

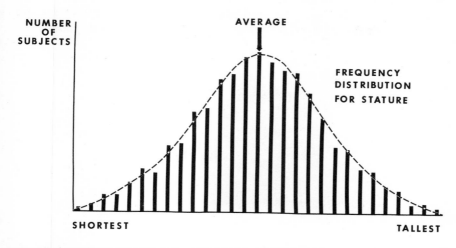

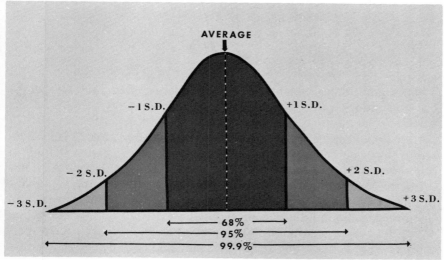

short or tall boys at the ends of the distribution.

Turning now to the curves for weight, the Iowa chart reveals that the median or fiftieth percentile is used to indicate the central tendency while the eighty-fourth and sixteenth percentiles are used as measures of dispersion. Why did the authors reject the average and the standard deviation in constructing the curves of growth for weight? Why are the measures of central tendency and dispersion different for weight than for height? The answers are to be found in the peculiar shape of the frequency distribution for weight. At any age, the distribution for weight is lopsided or skewed, with a long tail of heavy weights at the right. Heights are rather evenly distributed at any age, but weights are not. For this reason, the average and the standard deviation are not appropriate measures

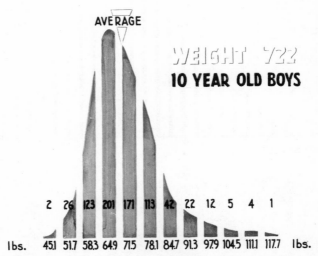

AVERAGE

WEIGHT 722
10 YEAR OLD BOYS

2 26 123 201 171 113 42 22 12 5 4 1

lbs.   45.1 51.7 58.3 64.9 71.5 78.1 84.7 91.3 97.9 104.5 111.1 117.7   lbs.

when constructing weight curves. Measures must be used which are not unduly influenced by a relatively few heavy weights in the sample. The authors of the Iowa curves, therefore, used the fiftieth percentile and the sixteenth and eighty-fourth percentiles. In calculating these percentiles, heads are counted rather than weights. In other words, all the weights are arranged in sequence from the lightest to the heaviest in order to determine the weights of those persons representing the point of central tendency (fiftieth percentile) and the limits of the middle two-thirds of the distribution (sixteenth and eighty-fourth percentiles). The sixteenth and eighty-fourth percentiles were deliberately selected as measures of dispersion because they correspond to the segment of the population incorporated between plus and minus one standard deviation. The percentile system escapes the distorting effect of the standard deviation when applied to a skewed distribution such as that for weight.

Having presented the manner of construction and the rationale of the Iowa curves, we can return to our original purpose—assessing a particular child's growth over a period of time. By using the Iowa graphs, it is possible to obtain a graphic picture of a child's physique, his status in height and weight, an evaluation of his progress, and an estimate of his nutritional status.

To illustrate the application of these graphs, we have plotted the heights and weights of three 6-year-old girls.

	Jane	Mary	Ann
Height (inches)	43	45	47
Weight (pounds)	40	45	50

From the pictures it is evident that the girls differ appreciably

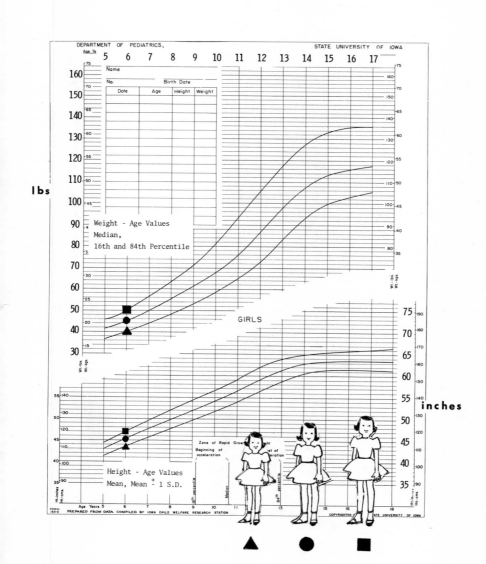

in stature and in weight. The *status* of each of these girls relative to the population may be assessed by determining their positions on the graphs. Jane is a small girl in height and weight; only 16 out of every 100 girls her age are shorter or lighter. Mary falls right on the average in height and on the median or fiftieth percentile in

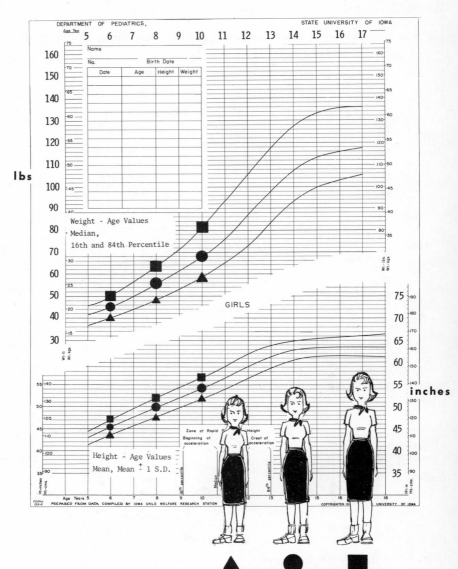

weight. Ann is the tallest and the heaviest of the three; only 16 out of every 100 girls her age will exceed her attainment in height and weight. Although these girls are markedly dissimilar in body size, they are comparable in one essential characteristic: they have the same physique. Each girl exhibits a similar relationship of weight to height. And, it is this relationship that determines physique.

Thus far we have shown how the Iowa curves are used to determine a child's status and physique from a single set of measurements. To judge the child's progress or nutritional adequacy, however, additional measurements at subsequent ages are needed. Accordingly, we have plotted the height-weight data for these girls at ages 8 and 10. Each girl has maintained the status and physique she exhibited at 6 years of age, and each has progressed according to her own potential, evidencing a progress consistent with her status at age 6. The graph also shows that the girls have maintained a constancy of physique (height-weight ratio) over the period studied. Since this is the case, we can infer an adequacy of diet and an absence of debilitating disease.

If the tallest girl, Ann, had sharply decreased in weight by age 10, so that her weight position placed her well below the median, we would be concerned about the inadequacy of her progress and would want to examine her nutritional intake and health history. In fact, this is one of the major contributions of a height-weight graph. Any radical shift suggests that a detailed examination of the child is required. The chart does not specify what is wrong but alerts us to the possible hazards of abrupt loss in status and inadequacy of progress.

The three girls are fictional cases. In actual life, children do not demonstrate such constancy of growth attainment. This fact is illustrated by plotting the height-weight data for two real children.

Looking first at the curve for stature of Gertrude Z, the illustration reveals a remarkable shift in status during the course of her growth. From 6 to 12 years of age she maintains a constant growth attainment, that of a relatively short girl whose height is exceeded by 85 percent of the population. At 12 years of age, however, she undergoes a dramatic spurt, raising her height well above the average. Thus, she has changed in status from a short girl to one whose height exceeds that of 50 percent of the population. Her weight curve is not nearly as dramatic; there is no abrupt shift in position during the course of her growth. Nevertheless, her gain in weight is steeper than that of the population; she moves from the sixteenth percentile at 5 years of age to the fiftieth percentile at 17 years of age. Thus, from the Iowa curves two significant facts may be deduced with reference to this child: she has maintained a fairly stable physique as evidenced by her height-weight relationship at each chronological age, and during the course of her development she has improved her status relative to the population.

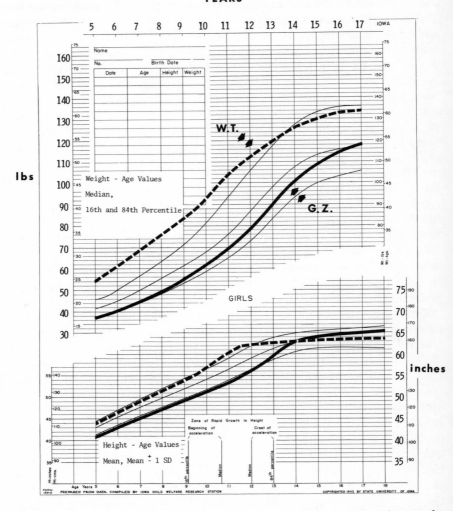

The curves for Wilma T. illustrate a contrasting picture of growth. During preadolescence she was taller than most girls her age. She stopped growing early, however, and her terminal height is that of an average-sized adult. Her statural growth reflects a loss in status relative to the population. On the other hand, her growth in weight reveals marked obesity until 10 years of age. During this period her weight attainments placed her at the hundredth percentile—at 8 and 9 years of age her curve almost runs off the graph. Subsequently, however, she exhibits an improved weight-to-height relationship reflected by the fact that her weight fell below the eighty-fourth percentile by the age of 14. Although she is making progress through yearly weight gains, her weight curve after age 11 is not as steep as that depicted for the population as a whole. Considering her weight curve in relation to height, we observe a tall and very heavy preadolescent developing into a stocky adult of average height.

These two girls exhibit radically different curves of growth in height and in weight. Further, both of them depart strikingly from the average curves depicted on the Iowa graph. Nevertheless, both are normal in the totality of their development. Wilma T., however, would be considered obese from 5 to 11 years of age. During this period her weight exceeded her height by more than two units. The distance from the mean to either plus or minus one standard deviation for height is one unit; correspondingly, the distance from the median to the sixteenth or eighty-fourth percentile for weight is also one unit. Although this girl was at the plus one standard deviation in stature, her weight curve exceeded her stature by more than two units. The authors of the Iowa curves regard a difference of this magnitude as an undesirable disparity. Whenever the height-weight relationship is more than two units removed, intensive examination of a child's physical well-being and nutritional intake is indicated. In the case of Wilma T., following such an examination at age 10, she was placed on a corrective diet, and a more desirable height-weight relationship was obtained.

The two cases illustrate a fundamental limitation in the utilization of normative curves of growth. It is not unusual to find that the individual child deviates markedly from these curves at different phases of his development. Since population curves lump all children together, they suppress individual differences. Consequently, early maturers cancel out late maturers, and the average curve tends to be bland and predictable in character. The individual, on the other hand, tends to show abrupt shifts, and the accelerating phase at adolescence is usually much steeper for the individual child than for the group as a whole.

Introducing Maturational Differences

An innovation in the construction of height-weight tables was introduced by Nancy Bayley in 1954, when she published separate curves of growth in height and weight tailored for children who mature at different rates. Recognizing that maturation controls growth, she devised curves for assessing boys and girls who mature skeletally at average, fast, and slow rates. Thus a child's size status is not assessed solely on the basis of chronological age but

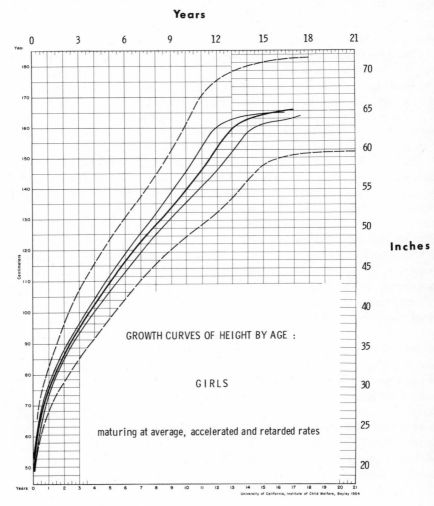

GROWTH CURVES OF HEIGHT BY AGE :

GIRLS

maturing at average, accelerated and retarded rates

University of California, Institute of Child Welfare, Bayley 1954

rather in term of maturational progress. To illustrate the contribution of skeletal maturation in appraising growth in height and weight, we shall examine the Bayley curves for girls.

On the Bayley charts, height is plotted in inches or centimeters on the vertical axis, while chronological age is indicated on the horizontal axis. In the body of the chart, five curves of growth are depicted. The middle three represent different rates of skeletal maturation: the upper curve, fast rate; the middle curve, average rate; and the lowest of the three, slow rate. What can be expected of the child who is maturing at a fast rate in contrast to the child maturing slowly? The rapidly maturing child, the one who is skeletally advanced, is taller than the average or slow child at each chronological age from early childhood through adolescence. For example, according to the chart, at 8 years of age a rapidly maturing girl is expected to be 2½ inches taller than a slowly maturing girl. The disparity is greatest at 12 years of age, when the rapidly maturing girl is about 6 inches taller than her slowly maturing contemporary. Rapidly maturing girls stop growing at an early age, while slowly maturing girls grow for a longer period of time and eventually catch up.

The broken curve at the top of the chart makes provision for those children who not only are fast maturers but are also constitutionally large. According to Bayley, these children will be tall adults and will attain their adult height 2 or 3 years before the average child. In other words, these children are taller at *every* age. Conversely, the broken curve at the bottom reflects the growth attainment of children who are constitutionally small and are maturing very slowly. They retain their smallness of stature into adulthood and never catch up to their peers.

Turning to the curves of growth in weight, we find a comparable series of five lines. Again, the three center curves represent, respectively, weight gains for rapid, average, and slow maturers. Also, the broken lines at the top and bottom represent children who are constitutionally large or small and maintain this pattern throughout development.

In order to use these charts as a diagnostic tool for assessing physique and nutritional status, it is necessary to relate height to weight. Bayley states that a child's weight curve should be in the same channel as his height curve; a child who is on the accelerated curve for stature should also be on the accelerated weight curve. Further, the child who is progressing satisfactorily should be evidencing comparable patterns of gain in height and weight. When

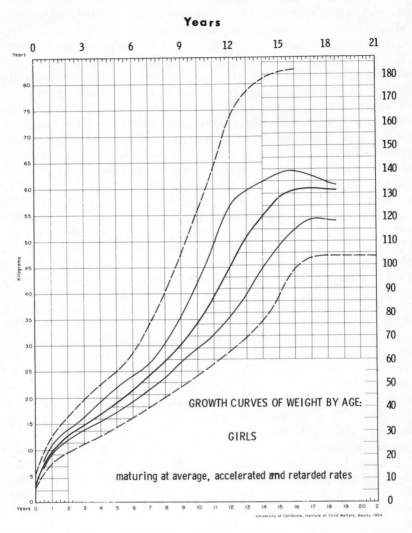

Years

GROWTH CURVES OF WEIGHT BY AGE:

GIRLS

maturing at average, accelerated and retarded rates

University of California, Institute of Child Welfare, Bayley 1954

there is an appreciable disparity in a child's relative positions on the height and weight curves, and therefore evidence of marked difference in annual height and weight gains, his nutritional status is suspect. If his weight position is markedly higher than his height position, the child is tending toward obesity; if the reverse is true the child may be demonstrating a nutritional deficiency. The prepubertal period, around 10 to 12 years of age, is a time when many children show excessive weight gains, causing a temporary discrepancy between the positions on the height and weight curves. The excess weight usually disappears as the child matures and reaches the statural growth spurt in adolescence. Of course, if the obesity does not disappear during adolescence, a special dietary plan may be called for.

By using the Bayley charts, unwarranted anxiety concerning growth can be dispelled, particularly for those children who appear to be unusually large or small. Individual differences in growth attainment can be better interpreted and understood. The value of the Bayley curves can be demonstrated by plotting and analyzing the growth data for an early maturing girl, the child previously illustrated on the Iowa curves.

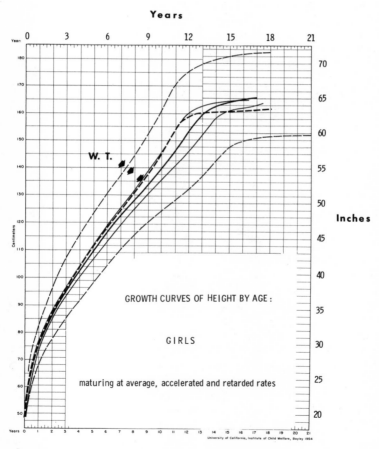

Wilma T.'s stature from 2 through 12 years of age approximates the curve of growth in height for accelerated girls. She is taller in childhood than her peers, and she stops growing earlier than more slowly developing girls. This is shown by the fact that her curve for stature plateaus at age 13. Despite her tallness in childhood, the Bayley curves tell us that she is not constitutionally a large person. Therefore, we can predict that she will not end up as a tall woman but rather that she is moving faster toward her terminal height than most girls. With these facts we can give her a more realistic picture of her eventual adult height. The shift in position suggested by the previous plotting of Wilma's growth data on the Iowa curve is a function of the early maturing pattern common to skeletally advanced children rather than the result of an actual loss in statural status.

An examination of Wilma's weight data from 2 through 8 years of

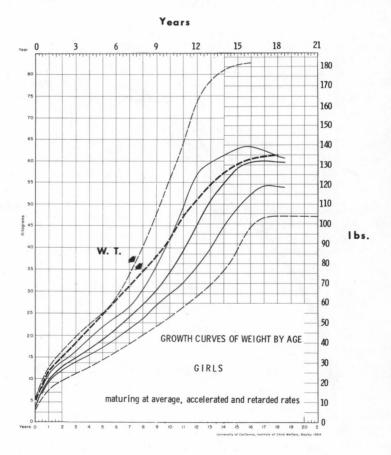

Years

W. T.

GROWTH CURVES OF WEIGHT BY AGE

GIRLS

maturing at average, accelerated and retarded rates

University of California, Institute of Child Welfare, Bayley 1954

age, plotted on the Bayley curves, gives the impression of a constitutionally large child. She approximates the uppermost curve of
the graph. However, since we know from the analysis of her stature
that she is a skeletally accelerated child, and not constitutionally
large, we arrive at a different interpretation. We conclude that she
is an obese child with a nutritional problem. The disparity between
her height and weight channels is an index of her obesity. As stated
earlier, a dietary plan introduced at 10 years of age established a
more desirable height-weight relationship. This pattern was maintained into young adulthood. Thus, the Bayley curves, utilizing
information on skeletal development, enable us to discriminate between children who are constitutionally large and those who are
simply accelerated in their maturation. By setting up different growth
expectations for children who develop at different rates, the Bayley
curves aid in reducing the bewildering complexity of individual variation.

The Pinching Approach

A radically different approach to the problem of assessing obesity
does not consider the relationship between height and weight at
all; in fact, it does not even consider stature a relevant variable.
The pinching approach attempts to assess the actual amount of
body fat. A simple procedure is used to measure the thickness of
fat under the skin. At any of the key surface areas of the body,
the skin and underlying fat are grasped between thumb and fore-
finger and pinched up into a fold; the thickness of the fold is then
measured with a caliper.

Two advocates of this technique, Drs. Seltzer and Mayer, object
to the use of height-weight tables on the grounds that a moderate
degree of overweight may be misinterpreted as an indication of
obesity. They state that a person may be overweight without being
obese. Certain athletes—for example, football linemen—are generally overweight but not obese; conversely, extremely sedentary
people can be obese without being overweight. Further, the height-
weight tables only permit inferences of obesity, but the pinching
approach gives a direct measure of adiposity, that is, body fatness.
This measure of obesity has been found to be virtually independent
of height.

Through research these investigators have determined that for
obese individuals the skin-fold at the back of the upper arm is most
representative of total body fatness and that no special advantage

is gained by measuring other areas of the body. It has also been determined that 50 percent of all body fat is found under the skin.

Based on a number of studies of skin-fold thickness, a table has been constructed giving the minimum obesity standards, measured in millimeters, for Caucasian American males and females up to the age of 30. The table listing these values for the back of the

Obesity Standards for Caucasian Americans

Age (years)	Minimum Triceps Skin Fold Thickness Indicating Obesity (millimeters)	
	Males	Females
5	12	14
6	12	15
7	13	16
8	14	17
9	15	18
10	16	20
11	17	21
12	18	22
13	18	23
14	17	23
15	16	24
16	15	25
17	14	26
18	15	27
19	15	27
20	16	28
21	17	28
22	18	28
23	18	28
24	19	28
25	20	29
26	20	29
27	21	29
28	22	29
29	22	29
30–50	23	30

Taken from: Seltzer and Mayer (1965). A simple criterion of obesity. *Postgraduate Medicine*, vol. 38, no. 2: A-101–A-107.

upper arm is reproduced here. Any skin-fold thickness in excess
of these values places the individual in the obese range. The final
listing on the table for age 30 also applies to older people up to the
age of 50. It is important to recognize that the minimum skin-fold
thicknesses are based on the plus one standard deviation for the population at each chronological age. By definition, then, 16 percent of
the American Caucasian population would exceed these figures and
would be designated as obese. After 30 years of age, unhappily, one
may expect to find a considerably greater percentage of individuals
who are obese.

When the data in the table are plotted in graphic form, an interesting sex difference in skin-fold thickness becomes apparent. While
a female undergoes a more or less continuous increase in skin-fold

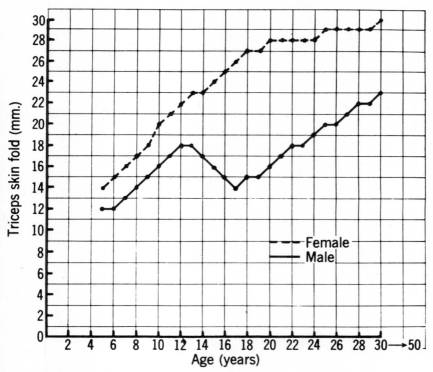

thickness from 5 to 30 years of age, the male evidences a sharp
decrease during adolescence, after which time an increase in skin-
fold thickness is resumed. We see in the arm, then, a phenomenon
we observed in studying differential tissue growth in the leg; the
male loses fat during adolescence, while the female exhibits an
appreciable increase.

The practical value of the skin-fold approach is that it offers a clear-cut criterion of obesity, a quick way of diagnosing it, and a simple method of measuring it in the individual.

We have reviewed a variety of devices for relating height and weight and have described how they are used to assess the developmental status of the individual. It is essential to recognize that whether height-weight tables or skin-fold measurements are used, these devices represent a distillation of statistical data drawn from population studies. At best, therefore, they are only screening mechanisms and cannot provide a definitive evaluation of a particular child. They cannot, for example, tell us that a child is suffering from a specific disease such as anemia. On the other hand, an abrupt shift or pronounced deviation from the normative pattern on any of the graphs should be viewed with concern, and an intensive examination of the individual should be undertaken. Invariably, additional information is needed to supplement height-weight data in order to obtain a comprehensive picture of a child's developmental and health status.

Chapter 9

THE TEETH

When an animal dies in the wild its flesh is often torn away by scavengers. The bones and teeth may be washed down the slopes of hills to adjacent rivers and streams and carried by the current until deposited on the river bottom or along the banks in some quiet eddy. Often the bones will become covered by fine layers of silt, and in time the accumulated weight of this deposited material will compress the lower layers into a compact sediment. Entombed within the sediment, the organic components of the bones will slowly decompose and eventually will be replaced by surrounding inorganic material. Much of the original mineral of the bones may also be removed and replaced by the sediment. Thus, in time, the original bone is supplanted by an exact replica of itself and becomes a fossil. The weight of the overlying sediment compresses the matrix around the bone into a sedimentary rock, and the bone becomes a record of past life.

Since the teeth are the hardest substance of the body, they are the portion least subject to decay and therefore most likely to be preserved in the fossil record. The teeth provide one of the most significant forms of evidence for the reconstruction of the evolutionary history of many animal groups. This is no less so for man. In large part, present-day conceptions of the origin and evolution of man from prehuman species are based on studies of fossil teeth.

The Evolution of the Teeth

Man is an old-world primate; his dentition reflects his affinity to the anthropoid apes of Asia and Africa. He has the same number of teeth as the old-world apes and monkeys—20 in the primary or baby set and 32 in the permanent set. In contrast, the monkeys of the new world have 36 teeth in their permanent set. Even man's tooth morphology, that is, the shape or form of the teeth, reflects his old-world origin. The paleontologist has been able to trace the evolutionary changes in the pattern of the cusps on the biting surfaces of the molar teeth from the generalized primitive apes who lived in Asia and Africa 25 million years ago to present-day man.

Man's dentition, in large part, is a reflection of the key events in his evolutionary history. Man's progenitors were generalized apes who spent a considerable portion of their lives in the trees; when they came down to the ground they moved about quadrupedally, that is, on all fours. They probably lived in small family groups consisting of a dominant male and several females with their offspring. Probably, too, these primates controlled a given territory through which they would wander searching for edible plants. Despite the fact that they were not predators and did not hunt other animals for food, they possessed massive faces and teeth. The large tusk-like canines or eye teeth of the male served him as a weapon of protection against predatory animals and also as a means of warding off contesting males who attempted to invade his domain and capture his females.

Early in his history man abandoned entirely the arboreal or tree existence of his ancestors and became a ground-dwelling creature. He developed a bipedal mode of locomotion and changed his mode of subsistence; no longer depending on fruits and plants, he developed into a hunter and predator.

The factors that caused a group of forest-dwelling apes to develop into predatory bipedal hunters are not known with certainty. A radical change in the environment probably precipitated this shift. It is believed that the tropical forests in which the prehuman apes lived began to dry up; the lush vegetation was replaced by savannah grasslands—open, treeless plains. Confronted by drastic changes in the environment, survival pressures favored those organisms which had the adaptive capacity to survive on the ground and sustain the chase for long periods of time, thereby wearing down the prey, which was swifter afoot over short distances. This transformation set off a new chain of evolutionary events.

Deprived of tropical vegetation, man became an omnivorous hunter, gathering foodstuffs and pursuing animal prey. He became dependent upon tools for survival and constructed and used primitive stone and bone implements. Thus, in South Africa, in association with the skeletal remains of the earliest man, the remnants of primitive stone tools and the bones of many animals were found. Included were the skulls and thighbones of many baboons. These remains comprised one of the earliest dramatic events in the history of forensic medicine. Dr. Raymond A. Dart has demonstrated the presence of double depressed fractures on the top of the skulls of these fossil baboons. He has shown that the condyles (the knobs at the bottom of the bone) of the baboon's thigh fit into the double depressions on the top of the baboon's skulls, thus documenting the fact that these bones were used as clubs.

The increasing dependence upon tools reduced the importance of teeth as weapons for combating predatory animals. Emphasis on tool development put a premium on intelligence and on brain development. Evolutionary history shows that the enlargement of brain size was accompanied by a progressive reduction in the size of the teeth and the face. The large cuspids of the anthropoids no

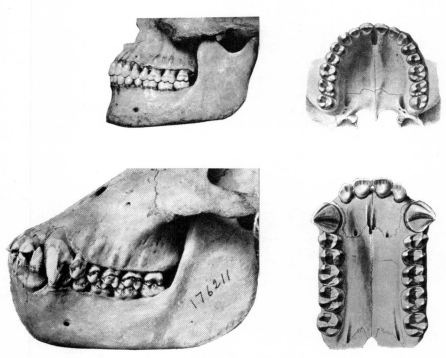

Upper — Man; Lower — Gorilla
Reproduced in true proportion

longer had a selective advantage, and these teeth underwent a radical reduction in size. In modern man, the cuspid or canine is so reduced in length that it barely projects beyond the adjacent teeth. The molars also became attenuated in size and form. In the apes and the earliest of human fossil men, the second molar was larger than the first and the third was larger than the second. During the evolutionary reduction of the teeth, the gradient of molar tooth size was reversed, so that in modern man the first molar is the largest and the third the smallest. In many individuals the third molar is imperfectly formed or does not appear at all.

Tools and culture became the surrogate for instinct and made possible man's control of his physical environment. Culture does for man what instinct does for the other animals. It is the medium of adaptive response to the environment. Man manipulates the environment rather than being genetically precommitted to it.

The Types and Functions of Teeth

Man is not as fortunate as the shark. Possessing successional teeth, the shark grows a replacement when one is lost. Further, the shark has a large number of undifferentiated teeth. Although the number may differ from shark to shark, they are all alike in form. In contrast, man is reduced to two sets of teeth and each set has a fixed number, which are specialized in type and function.

The first set of teeth, known as the primary dentition, begin to erupt around 6 months of age. By the time the child is approximately 2½ years of age, the entire set of 20 teeth will have erupted. The

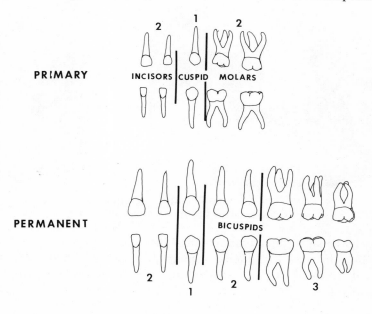

teeth are readily classified into three types: the incisors or cutting teeth, the canines or cuspids, and the molars. (The primary dentition is characterized by 8 incisors, 4 canines or cuspids, and 8 molars.) Since the right half of the face mirrors the left, and the same numbers and types of teeth are present in the upper jaw as in the lower, the dental apparatus is usually described in terms of one quadrant. Thus, the numbers and types of teeth in the primary set may be expressed as a formula: $(2\text{-}1\text{-}2)/(2\text{-}1\text{-}2) \times 2 = 20$.

The adult dentition introduces a new type of tooth not present in the primary set. This tooth is called a premolar or bicuspid since it has two cusps, one on the cheek side of the tooth and the other on the tongue side. Superficially, each bicuspid looks as if it were half a molar tooth, and in its function it serves the dual purpose of both cutting and grinding of food. In the carnivore, such as the dog, the premolar tooth looks more like a knife blade; working against the opposing member in the other jaw, it acts as a shears, which can be used to cut tendons and tough bits of meat. The permanent dentition differs, further, in that this set has 3 molars (rather than 2) in each quadrant. Thus, the adult dental formula is: $(2\text{-}1\text{-}2\text{-}3)/(2\text{-}1\text{-}2\text{-}3) \times 2 = 32)$.

Although much is made of the primary and permanent teeth as distinct sets, actually they are intimately related during development and eruption. During the replacement phase, parts of both sets are present in the mouth at the same time. Consequently, a person unfamiliar with tooth structure looking into a child's mouth might well be puzzled in attempting to identify the individual members of the two sets.

The Structure of the Teeth

All of the teeth are composed of the same tissues and exhibit the same fundamental structure, regardless of whether they are primary or permanent, incisor or molar. A tooth may be divided into two major parts: the crown and the root or roots. The crown is the shiny, white portion of the tooth visible in the mouth; the root is submerged in the gums and bones of the jaw, and little of it is evident in the young healthy mouth. As we grow older, our gums tend to recede, and progressively the root structure is exposed.

The molar depicted in the diagram shows the three-root structure characteristic of an upper molar tooth. (In contrast, the lower molar tooth has but two roots.) The crown is covered by an outer shell of enamel, the hardest tissue in the body. Encased within this shell is found the dentin, a somewhat softer tissue that forms most of

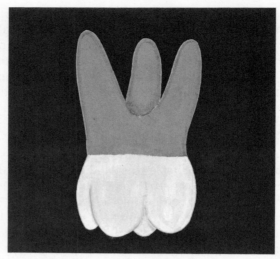

the bulk of the tooth. In contrast to the enamel, the dentin extends from the crown down to the tip of the root. Deep within the dentin at the core of the tooth is the pulp cavity, which contains a highly sensitive nerve and blood vessels that enter the tip of the root through a fine aperture. Therefore, the tooth pulp is in communication with the circulation and nerve supply of the body. The dentin

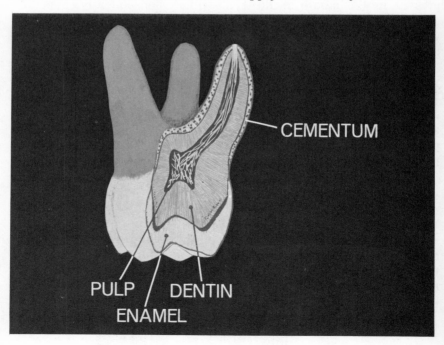

of the root is covered by a fairly tough material called cementum. Fibers from the gums penetrate into and through the cementum to anchor the tooth in its socket. Thus, enamel, dentin, pulp, and cementum comprise the tissue elements of the tooth.

The task of the enamel is to protect the dentin, and, in turn, the dentin protects the pulp. Once the pulp is exposed to air and to the saliva of the mouth, it becomes vulnerable to invasion by bacteria and to pain! The blood vessels and the nerves proper do not extend into the dentin or the enamel; therefore, these tissues in and of themselves are not sensitive to pain. A person may have a cavity in a tooth that exhibits considerable erosion of the dentin and yet have no awareness of this fact and little or no sensation. The American Indian who subsisted in large part on a harsh and gritty diet of corn often wore away the enamel covering on the biting surfaces of the teeth. In the making of clothing, Eskimo women cured hides by chewing them. This procedure rendered the hides soft and pliable. Not infrequently, by the time a woman was middle-aged she had worn away the crowns of her teeth, exposing considerable areas of the dentin.

The dentin, however, contains rods and nerve fibrils that permit the transmission of pressure and temperature changes to the pulp. When we drink unusually hot or cold liquids, we may experience temporary pain. Pilots flying at extremely high altitudes where the air pressure is low sometimes experience pain in teeth containing fillings. This pain is caused by air bubbles trapped under the fillings that expand in low atmospheric pressures. And you have noticed that when your dentist is using a high-speed air drill to excavate the decayed dentin in a cavity, he shoots a continuous spray of water on the tooth to cool it. Without this cooling mechanism, the intense heat built up by the high-speed drill would be transmitted to the pulp and might very well kill it.

It is important to care for the enamel and dentin because these tissues are incapable of regeneration. When a bone is broken or damaged, it has a mechanism for repair. Once a tooth is fully formed, however, the mechanism for laying down enamel disappears and the formative material of the dentin is retained as a less active layer of cells lining the pulp chamber. This layer cannot grow new dentin to replace the decayed material; consequently, the dentist must fill the cavity with amalgam or gold or some other restorative material. However, the layer of dentin-forming cells is stimulated or awakened into activity by an invading cavity. As if to shore up the tooth and protect the pulp from exposure, this mechanism forms new layers of dentin along the surface of the pulp. Unhappily, the rate of formation of new dentin on the inside of the tooth cannot keep pace with the erosion of the cavity, and unless the tooth is attended to it usually will be destroyed. The dentin formed in response to damage to the tooth is called *secondary dentin.* It will also be formed under a large filling or inlay placed in the tooth. Thus, we may experience unpleasant sensitivity in a tooth after a large filling has been placed in it. The unpleasantness may continue for a few weeks until the secondary dentin intervenes and provides a greater buffer between the filling and the pulp.

Growth of the Tooth

The growth of the tooth is a study in apposition. Like bone, or a tree trunk, or the scales of a fish, the crown and the root of the tooth grow as the result of the deposition of new layers of enamel and dentin on previously formed layers. The growth of the tooth follows a definite pattern or template, which is established early in its development. This template is in the form of a line roughly characterizing the shape of the crown. The line represents the dentino-enamel junction, the place where the enamel and dentin meet. The enamel matrix is deposited layer upon layer *outwardly* from the junction.

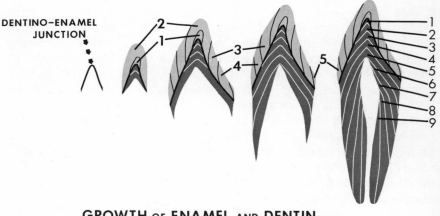

GROWTH of ENAMEL and DENTIN
(INCISOR)

Conversely, the dentin is deposited layer upon layer *inwardly,* away from the junction and toward the pulp cavity. When the final size of the enamel crown has been attained, the enamel-forming cells disappear and further formation of enamel is impossible. Dentin formation, however, exhibits one significant difference. The dentin-forming cells, unlike the enamel-forming cells, are not eliminated once the dentin of the tooth has been laid down. Rather, this tissue is reduced to a less active layer of cells lining the pulp cavity, as discussed previously.

In the young child, the pulp cavity is quite large relative to the amount of dentin already formed. For this reason, capping of teeth may be delayed for several years until the formation of the dentin has been completed.

The teeth, like the bones, record the vicissitudes of life. A child who is subjected to severe illness or metabolic disturbance during tooth formation will evidence rings in the enamel and dentin, comparable to the lines of arrested growth in the bones. Unlike the bones, however, the rings or lines formed in the teeth are permanent, because the tooth is not reconstituted in the manner characteristic of much of the bone. These rings are called incremental lines because they reveal the amount of enamel or dentin between two lines.

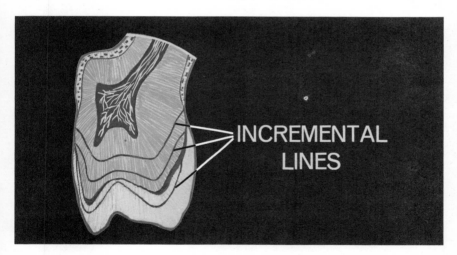

Investigators have found that even in healthy children a series of age-associated rings is regularly discernible. These rings occur at birth, at 3 months, 10 months, 2½ years and 5 years of age. In the opinion of dental histologists, these lines reflect periods of acute susceptibility resulting from systemic disturbance. The neonatal ring reflects the traumatic effects of birth and the change from a uniform intrauterine environment to the problems of extrauterine life. The fetus is a true parasite, deriving its mineral from the mother through the placenta. It is worth noting that dentin formed in utero is always of a more homogeneous texture and more finely mineralized than that laid down postnatally, no matter how well nourished the infant is subsequent to birth. The 3-month ring is believed to mark the site of tooth formation at the time that the fetal mineral reserve is exhausted. Henceforth, the infant is fully dependent on the mineral obtained through ingestion. The 10-month ring, the 2½-year ring, and the 5-year ring apparently represent critical periods of vulnerability.

The accompanying table (adapted from Massler, Schour, and Poncher, 1941) indicates the periods during childhood of good and poor calcification of the primary and permanent teeth.

Periods of Good and Poor Calcification of Primary and Permanent Teeth

Excellent	4th month in utero to birth
Very Poor	birth to about 10 months of age
Good	10 months to about 2½ years of age
Poor	about 2½ to 5 years of age
Fair	6 to 10 years of age
Poor	about 10 to 13 years of age

The teeth are not a calcium reservoir—despite the popular assumption that such conditions as pregnancy may drain the mother's teeth of mineral. For many decades it was assumed that pregnancy represented a potentially dangerous situation for the mother's teeth. This was reflected in the phrase "a tooth for every child," as if the fetus focused on a single tooth and robbed that tooth of its mineral content. It is physiologically impossible for the fetus to deplete the mineral content of the mother's teeth or of a given tooth. It is the skeleton which constitutes the calcium reservoir, and it is the skeleton in which the mineral content is in a state of flux and can be withdrawn or deposited. The fact that mineral can neither be added to the tooth nor withdrawn from it physiologically, subsequent to its formation, is the primary reason that dentists strongly advocate fluoridation of the public water supply.

It has been found that the inclusion of sodium fluoride molecules into the forming tooth results in a tissue that is more resistant to the decaying and eroding effects of acids in the oral cavity. It has also been determined that sodium fluoride lends stability to the crystalline structure of the minerals in bone and provides resistance against certain types of bone disease, such as osteoporosis.

The recommended concentration of fluoride is one part fluoride per one million parts of water—a safe, effective dosage. In some areas of the country, the water supply is naturally fluoridated and the concentration has been found to be as high as twelve parts per million without any injurious physiological effects on the population. In areas of very high concentration, however, the proportionally greater deposition in the teeth often creates a browning or mottling effect rather than the desirable white sheen of the enamel.

It has been found that a number of drugs and chemicals administered orally or through injection will be taken up by the bloodstream and deposited in the growing teeth and bones. Often these chemicals are visually apparent to the naked eye. Since they are incorporated in the portion of the tooth or bone that is forming at the time that the chemical is circulating in the bloodstream, the chemical acts as a "vital marker" indicating the site of growth of the dentin and enamel in the case of the tooth, and the depositing surfaces in the bone. Several of the common antibiotic drugs behave as vital markers. The group of antibiotics called the tetracyclines, of which aureomycin is the most well known, behaves in this manner and creates a brownish discoloration in the tooth if the drug is administered to very young children. Under ultraviolet light, the tetracycline band fluoresces a bright yellow. The sun provides sufficient ultraviolet light to make the band of tetracycline visible to the eye and aesthetically unpleasing.

Timing is the key factor in the use of sodium fluoride to create a tooth structure that is more resistant to caries. In other words, the sodium fluoride must be administered while the crowns of the teeth are forming. Conversely, it is during the formative period of the teeth that the use of the tetracyclines should be avoided.

Maturation of the Dentition

Two distinct maturational systems may be recognized during the development of the dentition: the calcification sequence and the eruption sequence. The calcification sequence refers to the sequence and timing of the mineralization of the 52 primary and permanent teeth. The eruption sequence refers to the order and timing of eruption of the teeth into the oral cavity. The two sequences overlap extensively in time. Calcification begins at 4 to 5 months in utero and extends through 15 to 20 years of age, when the roots of the third molar are completed. On the other hand, eruption begins at 6 months of age, when the first baby teeth appear in the oral cavity, and is usually completed when the third molar is evident in the mouth. In examining the maturation of the teeth, we shall see that the dental system exhibits a continuity of development beginning with the formation of the teeth in utero and extending through old age.

The intimate relationship of the primary and the permanent teeth during development is illustrated in the chart devised by Drs. Schour

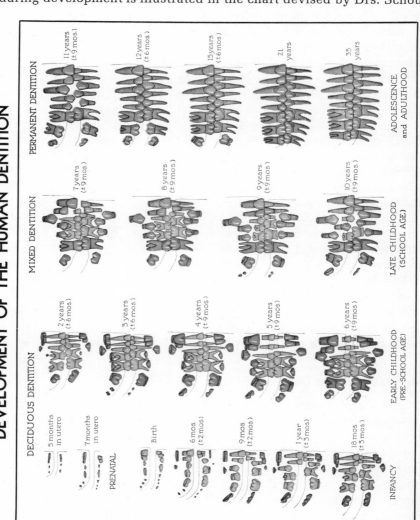

DEVELOPMENT OF THE HUMAN DENTITION

and Massler. Examination of this chart highlights several factors concerning the sequences of calcification and eruption.

1. The primary and the permanent teeth do not calcify in the same order. The accompanying table indicates the approximate time of onset of calcification in each of the primary and permanent teeth.

Chronology of Tooth Development

	Tooth	Beginning Calcification of the Crowns of the Teeth	Crown Completed	Age of Eruption
		(months in utero)	(age in months)	(months)
PRIMARY	Central Incisor	4–4½	1½–2½	6–8
PRIMARY	Lateral Incisor	4½	2½–3	8–10
PRIMARY	Cuspid	5	9	16–20
PRIMARY	First Molar	5	5½–6	12–16
PRIMARY	Second Molar	6	10–11	20–24
			Age in years	(years)
PERMANENT	First Molar	Birth	2½–3	6
		(age in months)		
PERMANENT	Central Incisor	3–4	4–5	7
PERMANENT	Lateral Incisor	10–12* / 3–4	4–5	8
PERMANENT	Cuspid	4–5	6–7	11
		(age in years)		
PERMANENT	First Bicuspid	1½–2	5–6	10
PERMANENT	Second Bicuspid	2–2½	6–7	11
PERMANENT	Second Molar	2½–3	7–8	12
PERMANENT	Third Molar	7–10	12–16	15+

*When appreciable differences occur between upper and lower teeth, their chronology is indicated separately.

Adapted from: Table 4, Schour and Massler (1940). Studies in tooth development: The growth pattern of human teeth. *The Journal of the American Dental Association,* vol. 27, no. 12: 1918–1931; Tables 1 and 2, Schour and Massler (1941). The development of the human dentition. *The Journal of the American Dental Association,* vol. 28, no. 7: 1153–1160.

2. The teeth do not erupt in a simple sequence from the front of the face to the back in either set.

3. The sequence of eruption is different in the two sets.

In the primary set, the order of eruption is as follows: central incisor, lateral incisor, first molar, cuspid, and, last, second molar. Assigning each tooth a number according to its position in the jaws from front to back and arranging the numbers according to the sequence of eruption, the order is 1, 2, 4, 3, 5. In contrast, the

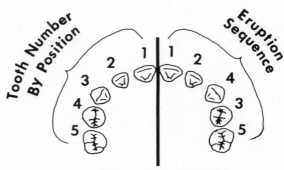

PRIMARY TEETH

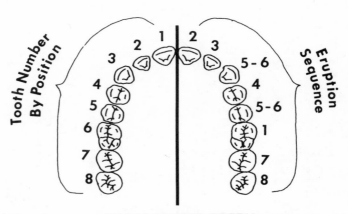

PERMANENT TEETH

sequence of eruption of the permanent teeth, using numbers corresponding to their position in the mouth, is 6, 1, 2, 4, 5—3, 7, 8.* Thus, the first molar (the 6-year molar) is the first tooth to erupt in the

---

*Some studies have found the following sequence to occur more frequently in the lower jaw: 6, 1, 2, 3, 4, 5, 7, 8.

permanent dentition. It is to be noted that at 6 years of age, prior to the loss of any of the deciduous teeth, the person has more teeth, visible and invisible, than at any other time in his life. By cutting away the outer plate of bone in the upper and lower jaws of the skull of a 4½-year-old child, it is possible to expose and visualize the developing permanent teeth in their crypts.

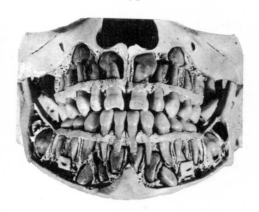

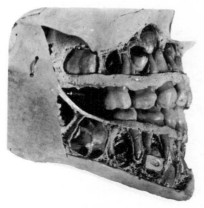

Further, the calcification sequence is rather analogous to the sequence of appearance of the centers of ossification in the skeleton and is fairly predictable from individual to individual. The eruption sequence, on the other hand, shows much greater individual variation both in the order and in the time of appearance of the individual teeth. The chart, therefore, must be taken to represent a generalized or modal individual. Consequently, the estimation of a person's chronological age based on his dental age (the stage of eruption of the teeth) can be very misleading.

With respect to the appearance of the dentition, the most critical period of eruption is that from 6 to 12 years of age—the period of mixed dentition. At this time both primary and permanent teeth are functionally present in the oral cavity. Too frequently, the primary teeth are thought of as expendable because they eventually will be exfoliated. These teeth, however, significantly influence the space relationship and the eruption of the permanent teeth. For example, if the second baby molar is lost prematurely, the first permanent molar, which erupts behind it, will tend to drift and tip forward. Thus, it will come to occupy the space that should be reserved for the second bicuspid, which normally erupts at about 11 years of age. It is imperative that this space be maintained. Proper dental care entails the use of a space maintainer to hold back the first permanent molar. This may simply consist of a metal bar anchored against the front of the first permanent molar and the back of the first primary molar.

The interaction of the two sets of teeth is reflected also in the eruption process. Prior to the loss of the primary tooth, its roots begin to resorb; according to some authorities this is caused by the pressure of the crown of the underlying permanent tooth. Ultimately, the entire root of the primary tooth is lost, and it is precariously held in the mouth solely by its attachment to the gums. In this state, it is possible, with sufficient wiggling, for a child to extract his own tooth. During the period of mixed dentition, then, while the roots of the permanent teeth are forming, those of the primary teeth are resorbing.

Further maturational changes of the dentition occur during adult life as a consequence of tooth wear. Attrition of the occlusal surface takes place where the upper and lower teeth meet. As a result, the contact points between adjacent teeth are worn away and spaces occur between them. This problem is offset by the tendency of the teeth to drift forward, thereby closing the spaces. Wearing down of the teeth is also compensated for by the tendency of the teeth to undergo further eruption. Frequently, however, continued eruption is not sufficient to replace the eroded tissue, and the individual experiences a closing down of the bite.

The maturation of the dentition and the correlated growth of the jaws are especially important for orthodontics, the branch of dentistry that is concerned with problems of malocclusion or poorly positioned teeth. It is not uncommon to find children with teeth too large for their jaws or children in whom there is disharmony in growth between the upper and lower jaws. These conditions pose special treatment problems, often requiring extraction of all 4 first bicuspids in order to provide sufficient space to fit the remaining teeth into an attractive, workable arch. Less frequently, extensive surgical procedures are required. In the event that the lower jaw is extremely long, resulting in a forward protrusion of the lower dentition, a mandibular resection may be necessary. This involves

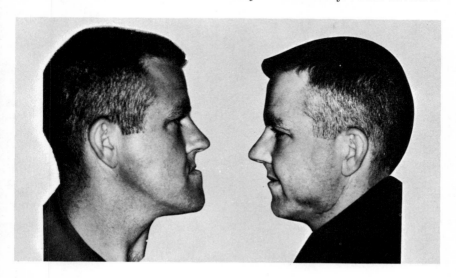

literally cutting segments out of the jaw and moving the front portion backward so that the lower teeth are in an effective relationship with the teeth of the upper jaw.

It is an ironic fact that civilized man has managed to undo 400 million years of evolutionary development of the dentition. The history of the vertebrates can actually be written in terms of the elaboration of effective dental apparatus and its importance for the survival of the organism. Animals with poor occlusion are unable to capture and ingest the foodstuffs necessary for survival and therefore are quickly eliminated. Man, on the other hand, can survive effectively with functionally inefficient teeth and even without teeth at all. His dependence upon technology has freed the dental apparatus from the evolutionary selective pressure that is operative in all nonhuman forms. Thus, we have come full cycle from the early Paleozoic times when the primitive vertebrates first developed teeth.

Chapter 10

THE INTEGRATION OF
GROWTH AND MATURATION

Throughout the preceding chapters, growth and maturation have been discussed as if they were completely separate attributes of development. While this separation is useful for purposes of distinguishing their characteristics, within the organism growth and maturation function in concert. The integration of these processes is accomplished through the endocrine system. In fact, the endocrine system is a necessary link in the control, direction, and synchronization of the totality of development.

The Endocrines

The endocrine glands control the metabolism and development of specific tissues, organs, and structures. They are commonly referred to as glands of internal secretion or ductless glands. Lacking a separate system of communication to the target organs and tissues under their influence, they pour their secretions directly into the bloodstream. The secretions proper are called hormones. Earlier, it may be remembered, we discussed the role of the parathyroid hormone (parathormone), which maintains the proper calcium level in the blood. Given a condition of sustained calcium deficiency in the diet, heightened parathormone secretion liberates calcium from the bone and restores it to the bloodstream.

The endocrine glands do not function autonomously but are actually a link in a larger chain of command. The present conception holds that the endocrine glands are under the control of the hypothalamus, a part of the brain situated adjacent to the base of the skull. It is believed that the hypothalamus sends out signals at the appropriate time to stimulate the functioning of a particular endocrine gland. These signals are transmitted via the pituitary, which serves as an intermediary. The hypothalamus, in turn, is under the control and influence of the hereditary material that contains all the coded instructions for the development of the organism. Thus, a particular aspect of development or function and the time at which it occurs is determined by a sequence involving a genetic blueprint, the hypothalamus, the pituitary, the specific endocrine gland, and, finally, the target organ or tissue.

Four of the endocrine glands have special relevance in the regulation of development. These glands are the anterior lobe of the pituitary, the thyroid, the adrenal cortex, and the gonads (the testes in the male and the ovaries in the female). The location of each of these is shown in the accompanying diagram.

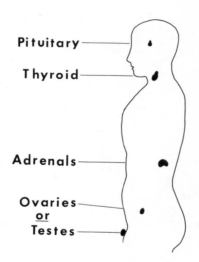

The pituitary is housed in a special bony pocket in the base of the skull, adjacent to the underside of the brain. The thyroids comprise paired structures at the front of the throat. The adrenals, also paired structures, are situated in the lower portion of the back, adjacent to the kidneys. The paired ovaries are contained within the lower part of the abdominal cavity, while the paired testes are suspended outside the abdominal cavity in the scrotum.

The anterior lobe of the pituitary has often been referred to as the master gland of the endocrine system because in addition to its own specific functions it also regulates the development and operation of the other endocrine glands in the body. The pituitary gland might even be considered an extension of the brain because in part it is derived from it and remains connected to it by a stalk. At least nine different functions have been documented for the anterior pituitary, each manifested through a specific hormone secreted by this organ.

One of the functions of the anterior lobe is the secretion of growth hormone (GH). This hormone is directly responsible for the growth of the body in size. In the event that a child's output of growth hormone is insufficient, he fails to achieve his full potential in stature. In extreme cases, at adulthood he is a midget with a stature comparable to that of a young child. The accompanying illustration shows an 8-year-old boy (B) who is suffering from an insufficiency

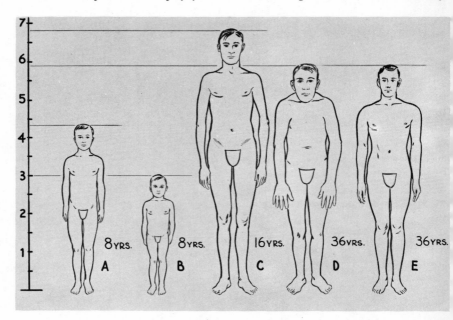

of growth hormone. Contrasting him with an average 8-year-old (A) dramatizes the growth-inhibiting effects encountered in *hypopituitarism*.

The pituitary midget tends to be normally proportioned, and his mental faculty is unaffected by this hormonal deficiency. He is a miniature of a normal individual of his age. It is difficult to detect pituitary dwarfing prior to 3 or 4 years of age. Apparently, until this time the growth of the body is mediated by other factors.

The opposite condition of pituitary malfunction, that is, an excessive production of growth hormone, is also found. In this event, the child undergoes exuberant growth and becomes inordinately tall. This condition of excessive statural attainment resulting from too much GH is called giantism. Earlier, in our discussion of stature, we noted that the legs grow during development at a disproportionately greater rate than does the head or the trunk. Contrasting a 16-year-old pituitary giant (C) with an average adult male (E) illustrates the retention of essentially normal proportions with the attainment of excessive height.

In the event that the person has an excess of GH output in adulthood, a different series of events occurs. The adult progressively acquires a distorted and misshapen appearance, as shown in the preceding illustration (D). This is because most of the centers of growth in the skeleton have lost their potential for further increase and cannot respond to an additional growth stimulus. Since growth in length of the long bones has been completed, no further statural growth can take place. Growth in thickness, however, is still possible, and under the stimulus of the growth hormone the bones become excessively thick and distorted in form. The chest cavity becomes tremendously enlarged and gorilla-like in its configuration. The hands and feet become massive. Perhaps the most dramatic changes occur in the head. The bones of the face become massive as a result of growth on their surface; the lips and facial tissue become heavy and coarse. The most grotesque effect is to be found in the mandible (the lower jaw). In contrast to the long bones, the hinge of the mandible (the condyle) comprises a site of bone growth that retains the potential for reactivation throughout life. Under the stimulus of the excessive growth hormone, the mandible exhibits an exuberant growth in length not matched by the upper

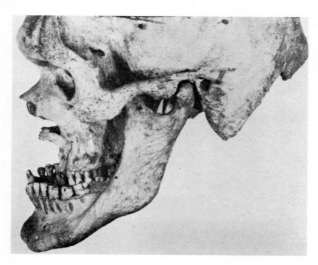

face. As a consequence, the lower jaw projects well out in front of the upper jaw, and the incisor teeth may be positioned a half-inch or more in front of the upper teeth.

This condition of excessive growth hormone in adulthood is called *acromegaly,* from the Greek, meaning literally "high and big." Acromegaly is a dramatic example of the significance of timing for development. In childhood, hyperpituitarism results in giantism, while the same condition in adulthood produces a grotesque and misshapen individual. Thus, the specific effects of hyperpituitarism are dependent upon the maturational level of the skeleton.

A little more than a decade ago, endocrinologists succeeded in extracting pure crystalline growth hormone from the human pituitary for use in the treatment of growth deficiency.* Although the number of children treated thus far is relatively small, the therapeutic use of growth hormone appears to hold great promise for stimulating growth in height. The goal, of course, is to intensify growth in height without excessively advancing skeletal maturation, since the early maturation of the skeleton would curtail the length of time remaining for the child to grow. Earlier we remarked that maturation controls growth and that the fusion of the epiphyses to the shafts of the long bones brings growth in length to an end.

Continuing the examination of the role of the anterior lobe of the pituitary, we find that it secretes a series of hormones specifically designed to induce the development of the other glands. One of these is thyrotropic hormone, which induces the normal development and functioning of the thyroid. The thyroid, in turn, secretes its own hormones, among which thyroxin is the most significant. Thyroxin controls body metabolism and maturation. In the classic situation, when a child is hypothyroid, that is, suffers from an insufficiency of thyroxin, he fails to mature. The child who is hypothyroid at birth is called a *cretin.* In the event that this deficiency is not diagnosed and endocrine therapy introduced, the maturation of the child is dramatically retarded, both physically and mentally. It was not uncommon in the past to find untreated cretins who at 13 years of age evidenced a level of maturation character-

---

*During the first week of 1970, the newspapers reported that biochemists had succeeded in synthesizing the human growth hormone. This complex task entailed connecting 134 amino acid units into a single chain in a very precise order. Although many months or even years may elapse before sufficient GH is manufactured, this breakthrough holds a great promise for treating thousands of children for whom the hormone is not now available. At the present time, GH must be extracted from the pituitary of hundreds of human cadavers in order to obtain enough in purified form to treat one or two patients.

istic of a normal 2-year-old child. Even the skeleton would exhibit the maturational features equivalent to those of a 2-year-old. The illustration shows the characteristic size and appearance of an untreated cretin (C), contrasted with a 13-year-old suffering from hypopituitarism (B), and a normal 13-year-old boy (A).

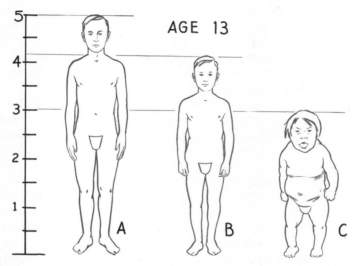

AGE 13

Many parents pride themselves in having a "good" baby. By this they mean that the child sleeps a lot, seldom cries, demands little attention, and evidences little irritability. Far from being a "good" baby, the young infant who is too quiet may be hypothyroid. The normal, healthy infant should and does demand attention; he becomes irritable when hungry, or when unfed, or if he is ignored for too long a period of time. Typically, the cretin or hypothyroid infant is sluggish, exhibits impaired circulation, has a cool, dry skin and poor muscle tone, and is given to constipation. He reveals retarded bone and tooth development, often exhibits a thick, protruding tongue, is mentally retarded, and fails to grow. During development, the untreated cretin acquires a disproportionately long trunk and short limbs, in contrast to the normal proportions of a child suffering from an insufficiency of growth hormone. Perhaps the principle characteristic of the hypothyroid child is infantilism, the retention of infantile characteristics into childhood. Hypothyroidism during the first 6 years of life is usually attended by mental retardation; the initiation of this state after 6 years results in a much less drastic effect on mental functioning. This is consistent with the fact that growth of the brain is early and rapid; 90 to 95 percent of the adult brain mass and organization is achieved by approximately 5 to 6 years of age.

Happily, hypothyroidism is an endocrine dysfunction that is amenable to treatment with thyroxin. The essential element in the picture is that the earlier the introduction of treatment, the better, since the child in whom treatment has been too long delayed may be incapable of making up significant amounts of arrested development. In extreme cases of hypothyroidism there also tends to be a deficiency of pituitary functioning; consequently, these children may suffer from an insufficiency of several hormones.

Children with hyperthyroidism resulting from an excess of thyroxin are behaviorally and physiologically the opposite of hypothyroid children, discussed above. These children exhibit a high rate of metabolic activity, causing nervous irritability, excessive movement, emotional instability, and tremor of the extremities. They have a high pulse rate and high blood pressure, as well as an enlarged thyroid; they appear exophthalmic, that is, their eyes bulge. While they may possess large appetites, because of their high metabolic consumption, nevertheless, they often exhibit a loss of weight. In contrast to the developmental deficiency of the hypothyroid child, the child with an excess of thyroid hormone does not undergo excessive development.

Two additional functions of the anterior lobe of the pituitary are intimately involved in the development of adolescence—the secretion of adrenocorticotropic hormone (more popularly known as ACTH) and gonadotropic hormone. The ACTH stimulates the development of the adrenal cortex, the outer shell of the adrenal glands. The importance of the adrenal cortex lies in the fact that it secretes androgens. Androgen is usually thought of as comprising a male sex hormone, but androgens are also liberated in the female. In fact, the development of pubic hair in the female results from the production of androgens by the adrenal cortex. The release of gonadotropins by the anterior pituitary results in the development and functioning of the ovaries and the testes. In turn, these glands produce estrogens, female sex hormones, and androgens, male sex hormones, respectively.

The phenomenon of adolescence is triggered by the measurable secretion of ACTH and gonadotropins into the bloodstream. Among the earliest events marking this dramatic period of metamorphosis is the prepubertal acceleration of growth in stature. This event, according to the available evidence, is unique to the primates and is somehow related to a protracted period of development. It is believed that the preadolescent acceleration in both sexes is a result of the heightened adrenal output. This assumption is supported by the observation that the administration of estrogen does not cause much acceleration of growth in sexually immature girls.

In both sexes, further, the androgenic output is responsible for the development of sexual hair (pubic hair), the sebaceous glands, and the musculature. At adolescence, both sexes evidence a significant increase in muscle mass. This is especially dramatic in the male, in whom the delicate contours of the child are transformed into the sharply delineated and angular features characterizing the masculine physique. Although the female also experiences an increase in muscle volume, the gain in the male is both relatively and absolutely greater. It may be remembered from the earlier discussion of differential tissue growth that at adolescence the male undergoes a relative loss in subcutaneous fatty tissue and a pronounced increase in muscle and bone. Conversely, in the female there is a relatively much greater increase in subcutaneous fatty tissue. At adolescence, then, the not too dissimilar physiques of the young boy and girl diverge, resulting in the obvious sexual dimorphism apparent at young adulthood.

Under the influence of the gonadotropic hormones, which are poured into the bloodstream concurrently with the heightened secretion of ACTH, the ovaries in the girl and the testes in the boy undergo a dramatic increase in size and mature to functional efficiency. With this event, the ovaries and testes secrete their respective sex hormones, estrogens in the female and testosterone in the male. The growth and maturation of the primary sex organs is accompanied by the development of the secondary sex characteristics. Most prominently, in the female, we note the pronounced curvature of the hips, the appearance of pubic and axillary hair, the appearance of protuberant breasts, and, of critical importance, the achievement of menarchy—the first menstruation. In the male, we observe the rapid enlargement of the penis and scrotum, the development of pubic and axillary hair, deepening of the voice, and incipient formation of facial hair.

The sexual hormones have an interesting retaliatory effect on the growth of the individual. These hormones operate as a feedback mechanism and bring about a terminus to growth. This effect is particularly dramatic in the female, in whom the statural curve of growth plateaus very abruptly once sexual maturity has been achieved. It is hypothesized that the sex hormones operate in one of two ways: either by suppressing the amount of growth-stimulating hormone secreted or by reducing the efficacy of the growth-stimulating hormone that is secreted. Since growth-stimulating hormone is present in the adult, the latter assumption is more likely. While the precise biological mode of interaction is still to be determined, we can state, nevertheless, that maturation controls growth. This relationship also emphasizes the significance of timing and duration of growth in different individuals. Thus, a girl who carries the genetic endowment that triggers her sexual maturity at an early age also tends to evidence exuberant growth in early childhood and to reach her adult stature at a relatively early age. The intimacy of the growth-maturation relationship is illustrated in the fact that the peak of the adolescent growth spurt in the girl appears in the year preceding that in which menarche occurs. In late-maturing

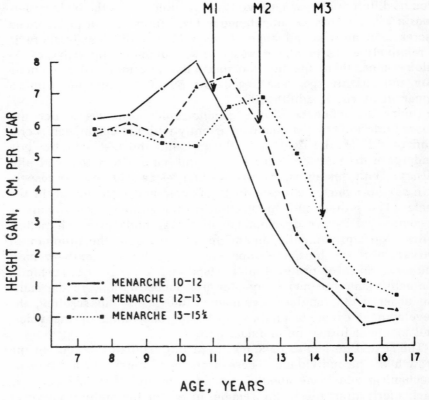

girls, however, an even greater time period elapses between the peak acceleration and the time of menarche. The inherited factor of timing determines the onset of maturational events and is responsible in large part for the tremendous variation in growth among individuals.

The fact that menarche regularly has been found to occur after girls have passed the peak of their height spurt can be used to reassure a tall girl who has begun menstruating that her growth is slowing down and will soon cease. The relationship of menarche to the rate of development of the breasts and the appearance of pubic hair is much more variable. Recent data indicate that menarche may not appear until the breasts are fully developed in some girls, while in others the breasts may be in a relatively immature stage at the time of the first menstruation. Comparable variation is found in the development of pubic hair.

Superimposed on the inherent variation is the further factor of environmental influence, which has been shown either to enhance or to retard the timing of development. Girls in technologically developed societies are reaching menarche at an earlier age than their mothers or grandmothers. This must certaintly be attributed to the more favorable environment in which they have developed.

The fact that a girl has begun menstruating does not necessarily mean that she is sexually mature in the sense that she is capable of bearing children. A number of anthropological investigations have revealed that in primitive societies where adolescent females may marry, the girls seldom have children until they have been married for several years. Ashley Montague has emphasized that the physiological requirements for reproduction are ovulation (development of a viable egg) and the maintenance of the fertilized egg within the uterus. He states that since these functions develop at some time subsequent to the first menstruation, conception usually is not possible during the period immediately after menarche. The period from menarche until the development of the hormones necessary to support the fertilized egg in the uterus is termed by Montague the period of adolescent sterility.

The endocrine control of sexual development, like all endocrine activity, may fall victim to malfunction, with consequent serious effects on the rate and pattern of sexual maturation of the child. The origin of the malfunction may occur at any one of the three links in the endocrine chain, that is, a disturbance of the hypothalamus, such as a brain tumor or lesion, damage to the pituitary or an aberration of the testes, ovaries, or adrenal glands. Damage to either the hypothalamus or the pituitary may inhibit or accelerate the time at which each of these structures is activated to send the necessary hormonal signals to the appropriate sex glands or the adrenals. Malfunctioning at any one of the three links in the endocrine chain may lead to precocity of sexual development or to sexual infantilism or even, in certain instances, to the superimposition of the secondary sex characteristics of the opposite sex on the child.

Perhaps the most dramatic illustration of premature endocrine activity is to be observed in cases of precocious puberty. While sexual precocity may be found in both sexes, statistical data show that it occurs twice as frequently in girls as in boys. Precocious sexual development can occur at an unbelievably early age. For example, Greulich and Pyle document the case of a girl who began menstruating at 7 months of age. By 1 year, 11 months, she evidenced incipient development of the breasts. Her growth in stature during early childhood far exceeded that of the normal girl, and by

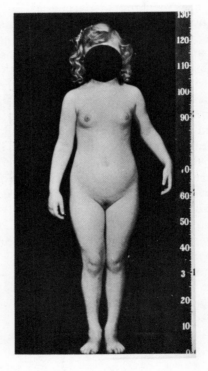

the time she was 5 years old, she was over 4 feet tall. Unfortunately, in this type of endocrine disfunction, growth usually ceases at a very early age and the child is abnormally short at adulthood. In this case, the girl ceased growing by the time she was 10, having reached a stature of only 4 feet 8 inches. Her skeletal development exhibited comparable acceleration. Thus, at 1 year, 11 months, she had attained a skeletal age of 5 years; at 3 years, 5 months, she had a skeletal age of 9 years, 6 months; and at 9 years, 11 months, the time at which she stopped growing, she had an assigned skeletal age of 14 years, 10 months. We see in this girl, then, the same intimate relationship between growth and maturation that is encountered in nomal development.

It is evident that accelerated sexual maturity does not stand out as a separate, isolated event within an otherwise normally developing child. The entire growth and maturation of the organism is progressively accelerated and terminated abruptly, to the detriment of the child. In precocious puberty, the gonadal hormones are secreted in sufficient quantities early enough to cause the fusion of the epiphyses in the long bones before normal adult stature can be achieved.

Precocious sexual puberty in the male exhibits a comparable picture of early rapid growth accompanied by accelerated skeletal maturation and a consequent early cessation of statural growth. The illustration from Wilkins offers a vivid example of this condition.

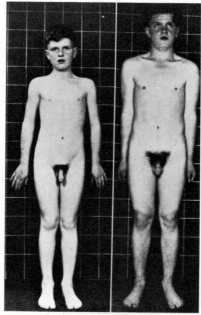

8 yrs. 1 mo.   10 yrs. 9 mos.

This boy acquired pubic hair at 7 years of age, his genitalia grew larger, and his voice deepened. He grew rapidly and became more muscular. At 10 years, 9 months, he had attained the height of the average boy of 13 years, 9 months. His skeletal maturation, on the other hand, was assessed at 18 years, 0 months, which means that his capacity for further growth was greatly diminished and he approached his adult stature. At this age, having attained a stature only comparable to that of a 13-year-old boy, it is obvious that this endocrine malfunction resulted in a statural achievement far short of the norm. Thus, the sexually precocious male tends to be characterized by shortness of stature, a heavily muscled physique, and a general physical appearance suggesting a maturity that belies his years.

Sexual precocity introduces special problems of psychological and social adjustment. In a sense, the problems confronting the child are analogous to those faced by any very early-maturing adolescent, who necessarily stands out as atypical within his peer group. Despite popular assumption, the sexually precocious child does not constitute a danger to the community. These children are as competent in regulating their sexual impulses as normal adolescents. Appropriate information and parental supervision can do much to alleviate the anxieties of the child and to facilitate an effective social adjustment.

Endocrinologists are also confronted with the antithesis of sexual precocity, that is, sexual infantilism—the failure of endocrine function leading to normal sexual maturity. This type of developmental anomaly may result from a hypothalamic disfunction, a deficiency in the pituitary, or a deficiency or absence of the ovaries or testes. Given any of these conditions, the child's normal sexual maturation will be interfered with; it may be severely retarded or even totally impaired. The extent of the damage determines the development of the secondary sex characteristics and the skeletal maturation of the child as well. Depending upon the type and severity of disfunction, endocrine therapy may be introduced to bring about the development of the secondary sex characteristics. Ordinarily, this program would not be initiated until the time when normal secondary sex changes would occur at about 12 or 13 years of age. If the deficiency is a result of failure of the pituitary to secrete gonadotropic hormones necessary for the development of the testes or ovaries, then the individual may be treated with gonadotropin to stimulate proper gonadal development.

In the event that the problem consists of a failure of normal embryonic differentiation of the gonads, or if the child has suffered prepubertal castration or degeneration of the gonads caused by disease, the administration of gonadotropins cannot be effective. This type of intervention is of no use, since the target endocrine glands have been destroyed. However, the administration of male or female sex hormone, normally secreted by the gonads, is of paramount importance. Without this therapy the child would fail to develop secondary sex characteristics and the attendant maturation of the skeleton. The epiphyses of the skeleton would remain open, and the child would continue to grow to an excessive height. As a result of differential growth, the legs would become relatively long, and the individual would take on eunuchoidal proportions.

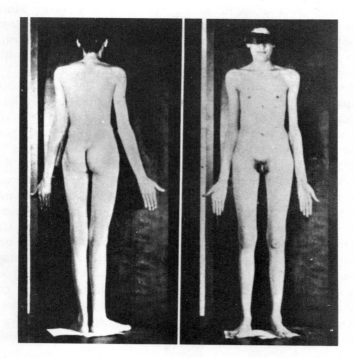

While sex hormone therapy, if administered at the appropriate time, may enable the individual to achieve a normal physical appearance, it cannot induce fertility, since the basic organs either are not present or are incapable of functioning.

Thus far we have discussed the effects of excessive or insufficient functioning of the ovaries and testes, and even the absence of these organs, on the development of the individual. It is also possible for a person to manifest traits of the opposite sex as a consequence of either faulty embryonic differentiation or postnatal malfunction of the endocrine glands related to the development of the secondary sex characteristics. This condition is known as *hermaphroditism,* a term taken from Greek mythology. The god Hermaphroditos was the son of Hermes and Aphrodite. As a consequence of uniting his body with the sea nymph, Salmacis, he took on the characteristics of both sexes.

A true hermaphrodite is a person in whom the primary sexual organs of both sexes have developed (ovaries and testes). Although imperfectly formed, many of the related reproductive organs of both sexes, as well as a vagina and a penis, may be present. Instances of true hermaphroditism, however, in which the individual possesses functional organs of both sexes, are extremely rare. More frequently, the individual possessing characteristics of the opposite sex is a pseudo-hermaphrodite, since the primary organs of only one sex are present. The secondary sex characteristics of the opposite sex, then, are simply superimposed on the individual.

Illustrative of a not uncommon anomaly is the true female who develops some of the secondary sex characteristics of the male. This condition is called *virilization* and may result from damage to the adrenal cortex. Thus, a tumor of the adrenals may result in excessive secretion of androgens, with severe consequences for the woman's appearance. Depending upon the severity of the damage, she may take on a masculine appearance and acquire an excessive growth of hair on the face, chest, and trunk. The clitoris may undergo hypertrophy and superficially resemble a penis; the voice may deepen, and excessive muscular development may occur. Usually, virilization is accompanied by a cessation of menstruation or menstruation becomes scanty and irregular. Further, the woman produces a high output of 17-ketosteroids (male sex hormone). Fortunately, the administration of cortisone has been found to cause a marked and permanent suppression of the 17-ketosteroid output in instances of virilization.

Cases of sexual transformation, which have received considerable notoriety in the newspapers, should not be confused with instances of endocrine pathology as such. Although surgery is used to correct conditions of true hermaphoditism in order to remove the organs of one sex and make the person fully of a single sex, this situation differs markedly from the well-publicized cases of sexual transformation. Most of the latter cases have a psychological rather than a biological origin. The individual is unable to accept his own sex identification and is incapable of functioning within it. Therefore, the person resorts to surgical intervention in an attempt to change his sex. More than 90 percent of the individuals desiring a sex transformation are men who would prefer to be women.

As a matter of fact, the person is not actually transformed into a biologically functioning member of the opposite sex; rather, he is made to resemble such a person. In the case of a male, he is castrated and the penis removed, and a vagina constructed surgically. Lacking the primary sex organs of the female (the ovaries as well as the uterus), he is incapable of producing female sex hormones (estrogens). Consequently, these hormones must be injected in massive quantities in order to induce the appearance of the female secondary sex characteristics. Under the influence of estrogens and surgery, the transformed male develops breasts, a lighter voice, a softer, rounded physique as a result of deposition of subcutaneous fatty tissue, and the absence of facial hair. The feminization of appearance can be maintained only by continuous administration of estrogens. Since the person is actually a castrated male and only displays a feminine appearance, he is obviously incapable of child bearing. Equipped with a female facade, he is supposedly qualified to play the sex role he desires.

In this chapter we have illustrated how the endocrine glands mediate the integration of growth and maturation. We have also stated that the endocrines do not function autonomously but are actually a link in a larger chain of command controlled by the genetic blueprint. In fact, all of the growth and maturational attributes that we have examined in this volume, as well as the endocrine mechanisms, are ultimately controlled and triggered by the succession of coded information contained in the DNA. The changes that occur during maturation and the pattern of growth of the different organs and tissues are reasonably well known and predictable; the differences in timing and rate of development, however, result in a dramatic variety of developmental expression. The genetically induced variation in timing and rate confronts us with a predictive challenge when we attempt to evaluate the developmental attainments of a particular child. At every point in his developmental career, the individual is a product of his inheritance expressed through a specific environment. Since the genetic blueprint is dependent upon the environment for its survival and fulfillment, it is the challenge of society to create an environment consonant with the biological nature of man.

References

General

Bayer, L. M., and N. Bayley (1959). *Growth Diagnosis*. Chicago: The University of Chicago Press.

Breckenridge, M. E., and E. L. Vincent (1960). *Child Development*, 4th ed. Philadelphia: W. B. Saunders Company.

Falkner, F. (1966). General considerations in human development. In *Human Development*, F. Falkner, editor. Philadelphia: W. B. Saunders Company.

Stott, L. H. (1967). *Child Development*. New York: Holt, Rinehart and Winston, Inc.

Tanner, J. M., editor (1960). *Human Growth*. Symposia of the Society for the Study of Human Biology, vol. III. Oxford: Pergamon Press Ltd.

Tanner, J. M., and G. R. Taylor (1965). *Growth*. New York: Time-Life Books.

Watson, E. H., and G. H. Lowrey (1967). *Growth and Development of Children*, 5th ed. Chicago: Year Book Medical Publishers, Inc.

Chapter 1

Greulich, W. W., and S. I. Pyle (1959). *Radiographic Atlas of Skeletal Development of the Hand and Wrist*, 2nd ed. Stanford, Calif.: Stanford University Press.

Harris, D. B., editor (1957). *The Concept of Development*. Minneapolis: The University of Minnesota Press.

Krogman, W. M. (1943). Principles of human growth. *Ciba Symposia*, vol. 5, nos. 1 and 2: 1458–1466.

Krogman, W. M. (1950). The concept of maturity from a morphological viewpoint. *Child Development*, vol. 21, no. 1: 25–32.

Martinez, C. P. (1960). Cranial deformations among the Guanes Indians of Colombia. *American Journal of Orthodontics*, vol. 46, no. 7: 539–543.

Scammon, R. E. (1930). The measurement of the body in childhood. In *The Measurement of Man*, by J. A. Harris, C. M. Jackson, D. G. Paterson, and R. E. Scammon. Minneapolis: The University of Minnesota Press.

Schour, I., and M. Massler (1943). Endocrines and dentistry. *The Journal of the American Dental Association*, vol. 30, no. 7: 595–603; no. 9: 763–773; no. 11: 943–950.

Todd, T. W. (1937). Orthodontic implications of physical constitution in the child. *International Journal of Orthodontia and Oral Surgery*, vol. 23, no. 8: 791–799.

Todd, T. W. (1937). *Atlas of Skeletal Maturation (Hand).* St. Louis: C. V. Mosby Company.

Chapter 2

Auerbach, C. (1956). *Genetics in the Atomic Age.* Fair Lawn, N.J.: Essential Books, Inc.

Dunn, L. C. (1959). *Heredity and Evolution in Human Populations.* Cambridge, Mass.: Harvard University Press.

Hunt, E. E. (1966). The developmental genetics of man. In *Human Development,* F. Falkner, editor. Philadelphia: W. B. Saunders Company.

McKusick, V. A. (1969). *Human Genetics,* 2nd ed. Englewood Cliffs, N.J.: Prentice-Hall, Inc.

Medawar, P. B. (1959). *The Future of Man.* New York: Basic Books, Inc.

Tanner, J. M. (1953). Inheritance of morphological and physiological traits. In *Clinical Genetics,* A. Sorsby, editor. London: Butterworth and Company Ltd.

Chapter 3

Dearborn, W. F., J. W. M. Rothney, and F. K. Shuttleworth (1938). Data on the growth of public school children (from the materials of the Harvard Growth Study). *Monographs of the Society for Research in Child Development,* vol. III, no. 1.

Garn, S. M., and Z. Shamir (1958). *Methods for Research in Human Growth.* Springfield, Ill.: Charles C Thomas.

Garn, S. M. (1971). Status of work accomplishments as of April 30, 1971, in relation to the Ten-State Nutrition Survey. Interim Progress Report 71-3 on Contract HSM-110-69-22. Center for Human Growth and Development, The University of Michigan, Ann Arbor, May 10, 1971.

Greulich, W. W. (1957). A comparison of the physical growth and development of American-born and native Japanese children. *American Journal of Physical Anthropology,* vol. 15, no. 4: 489–515.

Howells, W. W. (1960). The distribution of man. *Scientific American,* vol. 203, no. 3: 113–127.

Jackson, R. L., and H. G. Kelly (1945). Growth charts for use in pediatric practice. *The Journal of Pediatrics,* vol. 27, no. 3: 215–229.

Krogman, W. M. (1941). Growth of man. *Tabulae Biologicae,* vol. 20, The Haag.

Krogman, W. M. (1950). A handbook of the measurement and interpretation of height and weight in the growing child. *Monographs of the Society for Research in Child Development,* vol. XIII, no. 3.

Newman, M. T. (1960). Adaptations in the physique of American aborigines to nutritional factors. *Human Biology,* vol. 32, no. 3: 288–313.

Randall, F. E. (1947). Certain considerations of stature-weight relationships of male white Army separatees. Memorandum Report No. 16, Climatic Research Laboratory, Office of the Quartermaster General.

Redfield, J. E., and H. V. Meredith (1938). Changes in the stature and sitting height of preschool children in relation to rest in the recumbent position and activity following rest. *Child Development*, vol. 9, no. 3: 293–302.

Roberts, D. F. (1953). Body weight, race and climate. *American Journal of Physical Anthropology*, vol. 11, no. 4: 533–558.

Subrahmanyan, V., K. Joseph, T. R. Doraiswamy, M. Narayanarao, A. N. Sankaran, and M. Swaminathan (1957). The effect of a supplementary multipurpose food on the growth and nutritional status of schoolchildren. *British Journal of Nutrition*, vol. II, no. 4: 382–388.

Chapter 4

Bateman, N. (1954). Bone growth: A study of the grey-lethal and microphthalmic mutants of the mouse. *The Journal of Anatomy*, vol. 88, part 2: 212–262.

Boyd, E. (1941). *Outline of Physical Growth and Development*. Minneapolis: Burgess Publishing Company.

Cobb, W. M. (1952). Skeleton. In *Cowdry's Problems of Ageing*, 3rd ed., A. I. Lansing, editor. Baltimore: Williams and Wilkins Company.

Enlow, D. H. (1963). *Principles of Bone Remodeling*. Springfield, Ill.: Charles C Thomas.

Garn, S. M. (1970). *The Earlier Gain and the Later Loss of Cortical Bone in Nutritional Perspective*. Springfield, Ill.: Charles C Thomas.

Kahn, F. (1943). *Man in Structure and Function*. New York: Alfred A. Knopf.

Krogman, W. M. (1951). The scars of human evolution. *Scientific American*, vol. 185, no. 6: 54–57.

Mackay, H. *Skeletal Maturation* (chart). Rochester, N.Y.: Eastman Kodak Company.

Netter, F. H. (1953). Nervous system. *The Ciba Collection of Medical Illustrations*, vol. 1. Summit, N.J.: Ciba Pharmaceutical Products, Inc.

Pritchard, J. J. (1956). General anatomy and histology of bone. In *The Biochemistry and Physiology of Bone*, G. H. Bourne, editor. New York: Academic Press Inc.

Stratz, C. H. (1904). *Naturgeschichte des Menschen*. Stuttgart: Verlag von Ferdinand Enke.

Chapter 5

Acheson, R. M. (1966). Maturation of the skeleton. In *Human Development*, F. Falkner, editor. Philadelphia: W. B. Saunders Company.

Bayley, N., and S. R. Pinneau (1952). Tables for predicting adult height from skeletal age: Revised for use with the Greulich-Pyle hand standards. *The Journal of Pediatrics*, vol. 40, no. 4: 423–441.

Garn, S. M. (1968). Lines and bands of increased density: their implication to growth and development. *Medical Radiography and Photography*, vol. 44, no. 3: 58–89.

Greulich, W. W., and S. I. Pyle (1959). *Radiographic Atlas of Skeletal Development of the Hand and Wrist,* 2nd ed. Stanford, Calif.: Stanford University Press.

Loomis, W. F. (1970). Rickets. *Scientific American,* vol. 223, no. 6: 76–91.

McLean, F. C. (1955). Bone. *Scientific American,* vol. 192, no. 2: 84–89.

Shuttleworth, F. K. (1937). Sexual maturation and the physical growth of girls age six to nineteen. *Monographs of the Society for Research in Child Development,* vol. 2, no. 5 (Serial No. 12).

Simmons, K., and W. W. Greulich (1943). Menarcheal age and the height, weight, and skeletal age of girls age 7 to 17 years. *The Journal of Pediatrics,* vol. 22, no. 5: 518–548.

Tanner, J. M., and R. H. Whitehouse (1959). *Standards for Skeletal Maturity. Part I.* Paris: International Children's Centre.

Tanner, J. M., R. H. Whitehouse, and M. J. R. Healy (1962). *A New System for Estimating the Maturity of the Hand and Wrist, with Standards Derived from 2,600 Healthy British Children. Part II: The Scoring System.* Paris: International Children's Centre.

Weinmann, J. P., and H. Sicher (1955). *Bone and Bones,* 2nd ed. St. Louis: C. V. Mosby Company.

Chapter 6

Johanson, G. (1971). Age determinations from human teeth. *Odontologisk Revy,* vol. 22, supplement 21. CWK Gleerup, Lund.

Krogman, W. M. (1938). The skeleton talks. *Scientific American,* vol. 159, no. 2: 61–64.

Krogman, W. M. (1939). A guide to the identification of human skeletal material. *FBI Law Enforcement Bulletin,* August: 1–29.

Krogman, W. M. (1943). Role of the physical anthropologist in the identification of human skeletal remains. *FBI Law Enforcement Bulletin,* July–August: 17–40.

Krogman, W. M. (1962). *The Human Skeleton in Forensic Medicine.* Springfield, Ill.: Charles C Thomas.

McKern, T. W., and T. D. Stewart (1957). Skeletal age changes in young American males, analyzed from the standpoint of identification. Headquarters Quartermaster Research and Development Command, Technical Report EP-45, Natick, Mass.

Stewart, T. D., and M. Trotter, editors (1954). *Basic Readings on the Identification of Human Skeletons: Estimation of Age.* New York: Wenner-Gren Foundation for Anthropological Research, Inc.

Chapter 7

Boyd, E. (1941). *Outline of Physical Growth and Development.* Minneapolis: Burgess Publishing Company.

Keys, A., J. Brozek, A. Henschel, O. Mickelsen, and H. L. Taylor (1950). *The Biology of Human Starvation.* Minneapolis: The University of Minnesota Press.

Macy, I. G., and H. J. Kelly (1957). *Chemical Anthropology*. Chicago: The University of Chicago Press.

Owen, G. M., and J. Brozek (1966). Influence of age, sex and nutrition on body composition during childhood and adolescence. In *Human Development*, F. Falkner, editor. Philadelphia: W. B. Saunders Company.

Reynolds, E. L., and P. Grote (1948). Sex differences in the distribution of tissue components in the human leg from birth to maturity. *The Anatomical Record*, vol. 102, no. 1: 45–53.

Scammon, R. E. (1930). The measurement of the body in childhood. In *The Measurement of Man*, by J. A. Harris, C. M. Jackson, D. G. Paterson, and R. E. Scammon. Minneapolis: The University of Minnesota Press.

Stuart, H. C., P. Hill, and C. Shaw (1940). Studies from the Center for Research in Child Health and Human Development, School of Public Health, Harvard University. III. The growth of bone, muscle and overlying tissues as revealed by studies of roentgenograms of the leg area. *Monographs of the Society for Research in Child Development*, vol. V, no. 3.

Chapter 8

Baldwin, B. T., and T. D. Wood (1923). Weight-height-age tables. Tables for boys and girls of school age. *Mother and Child*, supplement to July issue. Washington: The American Child Health Association.

Bayley, N. (1956). Growth curves of height and weight by age for boys and girls, scaled according to physical maturity. *The Journal of Pediatrics*, vol. 48, no. 2: 187–194.

Jackson, R. L., and H. G. Kelly (1945). Growth charts for use in pediatric practice. *The Journal of Pediatrics*, vol. 27, no. 3: 215–229.

Seltzer, C. C., and J. Mayer (1965). A simple criterion of obesity. *Postgraduate Medicine*, vol. 38, no. 2: A-101–A-107.

Chapter 9

Colbert, E. H. (1961). *Evolution of the Vertebrates*. New York: Science Editors, Inc.

Graber, T. M. (1966). Craniofacial and dentitional development. In *Human Development*, F. Falkner, editor. Philadelphia: W. B. Saunders Company.

Gregory, W. K. (1922). *The Origin and Evolution of the Human Dentition*. Baltimore: Williams and Wilkins Company.

Gregory, W. K. (1965). *Our Face From Fish to Man*. New York: Capricorn Books.

Hellman, M. (1941). Factors influencing occlusion. In *Development of Occlusion*, by W. K. Gregory, B. H. Broadbent, and M. Hellman. Philadelphia: The University of Pennsylvania Press.

Hurme, V. O. (1949). Ranges of normalcy in the eruption of permanent teeth. *Journal of Dentistry for Children*, vol. 16, 2nd Quarter: 11–15.

Lasker, G. W. (1961). *The Evolution of Man*. New York: Holt, Rinehart and Winston, Inc.

Massler, M., I. Schour, and H. G. Poncher (1941). Developmental pattern of the child as reflected in the calcification pattern of the teeth. *American Journal of Diseases of Children,* vol. 62, no. 1: 33–67.

Massler, M., and I. Schour (1958). *Atlas of the Mouth,* 2nd ed. Chicago: The American Dental Association.

Schour, I. (1938). Calcium metabolism and teeth. *The Journal of the American Medical Association,* vol. 110, no. 12: 870–877.

Schour, I., and M. Massler (1940). Studies in tooth development: The growth pattern of human teeth. *The Journal of the American Dental Association,* vol. 27, no. 11: 1778–1793; no. 12: 1918–1931.

Schour, I., and M. Massler (1941). The development of the human dentition. *The Journal of the American Dental Association,* vol. 28, no. 7: 1153–1160.

Chapter 10

Andersen, H. (1966). The influence of hormones on human development. In *Human Development,* F. Falkner, editor. Philadelphia: W. B. Saunders Company.

Greulich, W. W., and S. I. Pyle (1959). *Radiographic Atlas of Skeletal Development of the Hand and Wrist,* 2nd ed. Stanford, Calif.: Stanford University Press.

Henry, N. B., editor (1944). *Adolescence. The Forty-Third Yearbook of the National Society for the Study of Education, Part I.* Chicago: The National Society for the Study of Education.

Kaplan, S. A. (1964). *Growth Disorders in Children and Adolescents.* Springfield, Ill.: Charles C Thomas.

Marshall, W. A., and J. M. Tanner (1969). Variations in pattern of pubertal changes in girls. *Archives of Diseases in Childhood,* vol. 44, no. 235: 291–303.

Montague, A. (1963). *Anthropology and Human Nature.* New York: McGraw-Hill Book Company, Inc.

Schour, I., and M. Massler (1943). Endocrines and dentistry. *The Journal of the American Dental Association,* vol. 30, no. 7: 595–603; no. 9: 763–773; no. 11: 943–950.

Stuart, H. C. (1946). Normal growth and development during adolescence. *The New England Journal of Medicine,* vol. 234, May 16: 666–672; May 23: 693–700; May 30: 732–738.

Tanner, J. M. (1962). *Growth at Adolescence,* 2nd ed. Oxford: Blackwell Scientific Publications.

Weinmann, J. P., and H. Sicher (1955). *Bone and Bones,* 2nd ed. St. Louis: C. V. Mosby Company.

Wilkins, L. (1957). *The Diagnosis and Treatment of Endocrine Diseases in Childhood,* 2nd ed. Springfield, Ill.: Charles C Thomas.